PLANT GROWTH REGULATORS

Comprising papers (with discussions) read at a Joint Symposium, organised by the Pesticides Group of the Society of Chemical Industry and the Phytochemical Society, held on 8th and 9th January, 1968, at the School of Pharmacy, Brunswick Square, London, W.C.1

S.C.I. MONOGRAPH No. 31

SOCIETY OF CHEMICAL INDUSTRY
14 BELGRAVE SQUARE, LONDON, S.W.1
1968

Chairman of the Pesticides Group:
H. HAYHURST, Esq.

Chairman of the Phytochemical Group:
Prof. W. D. OLLIS

Organising Committee:
Dr. M. B. GREEN (Hon. Sec., Pesticides Group)
Dr. J. B. HARBORNE (Hon. Sec., Phytochemical Society)
Dr. M. ELLIOTT (Hon. Treasurer, Pesticides Group)

Hon. Recorder:
Dr. D. L. COLINESE (Pesticides Group)

Made and printed in England by
STAPLES PRINTERS LIMITED
at their Rochester, Kent, establishment

CONTENTS

FIRST SESSION

SECOND SESSION

THIRD SESSION

FIRST SESSION

Chairman: Prof. W. D. Ollis
(Chairman of the Phytochemical Society)

CHEMICAL CONTROL OF PLANT GROWTH

By R. L. WAIN

(Wye College, University of London)

IT is now well established that plant growth is under the control of chemicals elaborated by the plant itself. Three important groups of these endogenous growth-regulating compounds are now recognised: the auxins, the gibberellins and the cytokinins.

Auxins

The demonstration by Went[1] in 1926 that oat seedlings contained a diffusible substance which would promote their growth was the first clear indication that a growth hormone occurs in plants. The discovery eight years later by Kögl[2] that indole-3-acetic acid (IAA) is capable of promoting the elongation of plant cells, focused considerable attention on this compound, which is now recognised to be the most important of the auxins. It is worth recording that IAA had been found in human urine over 50 years before[3] though it was not synthesised[4] until 1904. IAA is widely distributed in plants, especially in actively growing tissues, where it appears to be synthesised from tryptophan by a series of enzyme-controlled reactions.[5] The amounts of endogenous IAA are extremely small and are controlled at the physiological levels required for normal growth by the capacity of the plant to biosynthesise the compound, to destroy it by the action of IAA oxidase,[6] and to conjugate the molecule with such compounds as aspartic acid,[7] thereby removing its activity as an auxin. IAA might also be destroyed in tissues exposed to light by a photo-oxidation mechanism.[8]

Auxin-induced effects in plants include increase in stem length due to promotion of cell elongation, inhibition of root growth, production of adventitious roots, inhibition of leaf and fruit fall and the setting of fruit in absence of pollination. The effect of IAA in promoting the inward curvature of split stem segments of etiolated peas is shown in Fig. 1 which illustrates the pea curvature test for auxin activity.

Since the discovery of IAA as an auxin, similar activity has been found in many synthetic compounds, and some of these have become of great agricultural importance as selective herbicides.

FIG. 1. *Pea curvature test for auxin activity*
Segments of etiolated pea stem split for two-thirds of their length bend outward in water (left) but inward in a solution of a growth substance (right)

Most of the synthetic auxins, e.g. the aryl- and aryloxy-alkanecarboxylic acids and the arylcarboxylic acids, possess an unsaturated ring and an acidic grouping. However, an aromatic or heterocyclic ring system would not appear to be essential since compounds of the type:

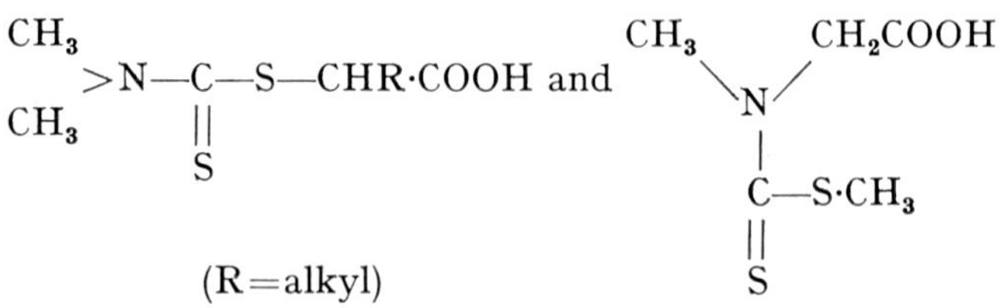

(R=alkyl)

are also active.[9,10] Other unrelated compounds, e.g. trichloroacetic acid and ethylenediaminetetraacetic acid show slight activity in the wheat cylinder test. Since this activity only becomes evident at concentrations just below those which produce toxic effects in the tissues used for bio-assay, these compounds should not be regarded as true auxins. Their apparent activity could arise from damaging effects on the cell wall which lead to changes in its plasticity.[11,12] When chlorine is substituted into the 2- and 4-positions of phenoxyacetic acid, the resulting compound (2,4-D) is highly active. With these substituents in the 2,6-positions, however, the compound is almost inactive. By contrast, 2,6-dichloro-phenylacetic and benzoic acids are appreciably active.[13,14] All of these groups of auxins find important uses as selective herbicides, and much work has been carried out on structure/activity relationships and their mode of action.

Gibberellins

The second group of plant growth hormones contains the gibberellins. This field of research began with the discovery by Kurosawa[15] in 1926 that the cell-free sterile filtrates of a fungus *Gibberella fujikuroi*, which causes pale spindly growth of rice, gave a marked growth stimulation when applied to seedlings of rice and grasses. At least 23 gibberellins have now been characterised; they differ only slightly in structure from the well-known gibberellic acid, GA_3, which is produced commercially from fungal cultures.

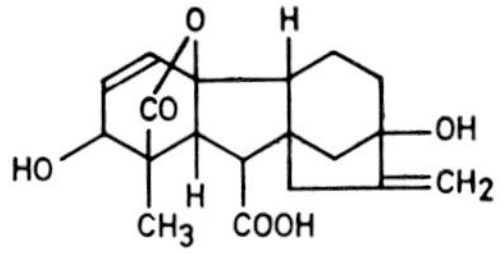

Gibberellic acid GA_3

There are now many reports showing that gibberellins occur naturally in plants,[16] and they probably operate closely with auxins in promoting cell extension growth. What was probably one of the first demonstrations of gibberellin activity in plant-extracts was made in 1951 by Mitchell. The effect of treating 'Black Valentine' bean seedlings with extracts of the immature seeds, although not recognised as such, was a 'gibberellin' effect and one which Mitchell was able to produce on several plants from the extract of a single seed, so potent was the active chemical present.[17] Gibberellins have since been isolated from bean seeds.[18,19] Gibberellins can produce considerable increases in the size of dwarf varieties of plants[20] and, in certain seeds, they play a vital role by promoting synthesis of hydrolytic and proteolytic enzymes upon which germination and seedling establishment depend.

Cytokinins

Both auxins and gibberellins are substances which promote cell enlargement. The other fundamental process which is concerned in plant growth is cell division, by which new cells are produced. Naturally occurring compounds which can influence this process are known as cytokinins. The first compound possessing cytokinin activity to be discovered was 6-furfuryl-adenine (kinetin), isolated in 1955 from autoclaved herring sperm DNA by Skoog and his co-workers.[21,22] Although cytokinins have been detected in extracts of immature seeds and many plant tissues,[23] it was not until 1964 that the isolation and identification of a naturally occurring cytokinin was reported. This compound, which was obtained from sweet-corn extracts by Letham,[24] was given the name zeatin. It was identified[25] and synthesised,[26] and like most other cytokinins it is a 6-substituted adenine derivative (6-(4-hydroxy-3-methylbut-2-enyl)aminopurine):

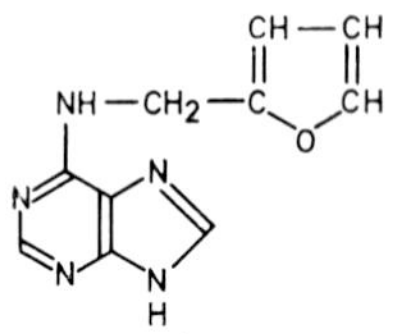

Kinetin

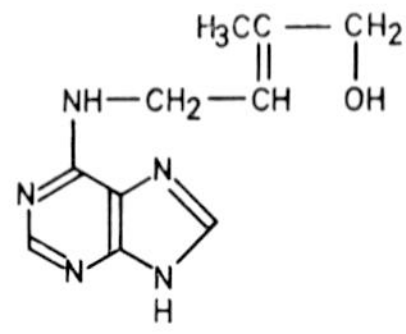

Zeatin

Cytokinins also have the property of inhibiting senescence; a detached radish leaf, for example, spot-treated with kinetin solution, remains green in the treated area when the rest of the leaf becomes yellow.[27] Furthermore, it has been shown[28] that a ^{14}C-labelled amino acid applied to the non-treated area of the leaf migrates to the site of cytokinin treatment; this indicates that the chemical is promoting normal metabolism at the expense of the rest of the senescing tissues.

Many 6-substituted adenine derivatives have been tested for phytokinin activity,[29] and a number of substituted ureas have recently been shown to be active.[30] Rothwell & Wright have synthesised and examined a number of adenine derivatives which have structural resemblances to kinetin:

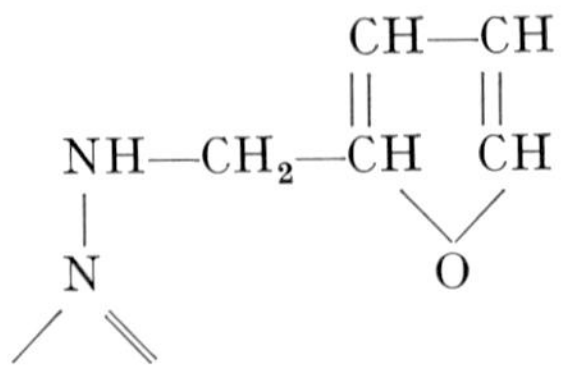

Kinetin

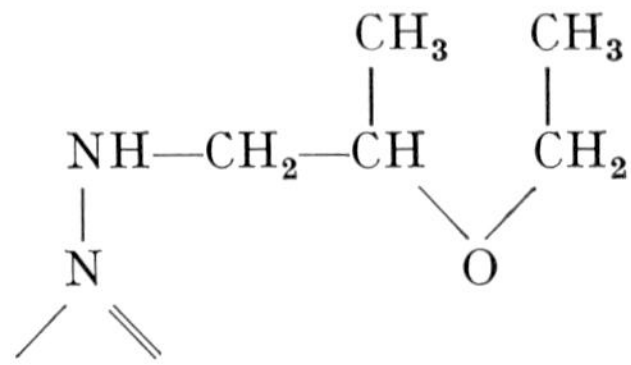

Ethoxypropyl derivative

A range of such compounds was tested and four were found to show activities of the same order as kinetin.[29] In the tobacco pith test of Murashige & Skoog,[30] it was found that although these compounds were not active at low concentrations at which kinetin and benzyladenine were still active, some of them were the more active at the higher concentrations.

Natural cytokinins operate with auxins and gibberellins in the hormonal complex which controls growth and development. In an attempt to determine whether in the growth of the wheat coleoptile, these operate together or in sequence, Wright made a study of the changes in the protein (and enzyme) pattern at different stages of growth.[31] Protein fractions were taken from coleoptiles at all growth stages and anti-serum was raised from the combined fractions in the rabbit. Using a technique involving the diffusion, in an agar gel, of protein extracts of coleoptiles of different ages and their antibodies, Wright was able to show that qualitative changes in protein occurred during growth and cellular differentiation of the wheat coleoptile. It was logical therefore to expect the responses of the coleoptiles to gibberellin, kinetin and IAA to vary at different stages of growth. This was found to be the case; indeed, Wright found clear indications that these substances exert their growth effects by acting in a well-defined sequence.[32] The first early phase of development was one of cell enlargement (approximately 0–30 hours after sowing) influenced mainly by GA, the second (approximately 30–60 hours after sowing) in which many of the cells were undergoing cell division was influenced mainly by kinetin. The third phase was a final period of cell enlargement (approximately 60–120 hours after sowing) which was promoted mainly by IAA. It has since been postulated that these three classes of growth regulators may operate in a similar sequential manner in the growth of fruit.[33] Furthermore, Nitsch has shown that tobacco pith tissue grown in a nutrient medium under sterile conditions also appears to have a sequential requirement for GA, kinetin and IAA.[34]

Abscisic acid

For some years it has been suspected that naturally occurring inhibitory compounds might operate in the chemical control of plant growth, and it is now accepted that such compounds (dormins) play an important role in controlling dormancy of buds.[35] At the Fifth International Conference on Plant Growth Regulators held in 1963, Addicott *et al.*,[36] Wareing *et al.*[35] and Rothwell & Wain[37] reported on the progress made in the extraction, isolation and identification of natural growth inhibitors present in young cotton fruits, sycamore leaves and yellow lupin pods, respectively. Shortly afterwards, the cotton fruit inhibitor was isolated and identified as 3-methyl-5-(1-hydroxy-4-oxo-2,6,6-trimethyl-2-cyclo-hexene-1-yl)-*cis, trans*-2, 4-pentadienoic acid.[38] This new plant growth regulator, which has been shown to inhibit the action of auxins, gibberellins and kinins, has been given the name abscisic acid:

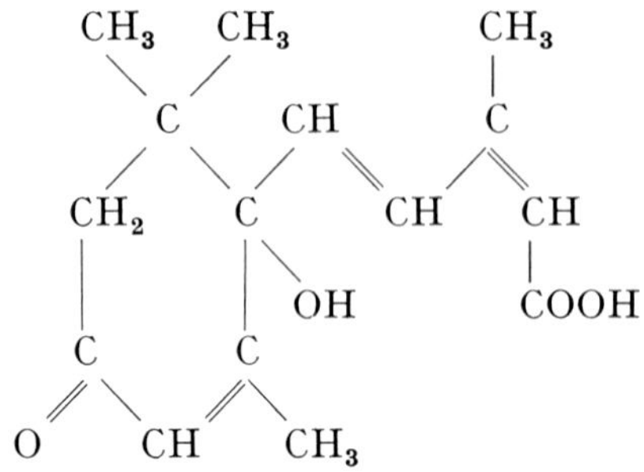

Abscisic acid

It has been identified as the inhibitor present in sycamore leaves and its structure has been established by synthesis.[39] Abscisic acid is also present in yellow lupin, is the dormancy-inducing inhibitor in birch leaves,[40] and the germination inhibitor in rose hips,[41] and it is present in many other plants. Taylor & Smith have described the production of seed germination inhibitors by the exposure to light of certain carotenoids—compounds which, it will be noted, show structural similarities to abscisic acid.[42]

Ethylene

This brief survey of naturally occurring compounds which operate in the regulation of plant growth must include reference to ethylene. This simple molecule has long been known to be highly active as a growth substance and there is abundant evidence that it is biosynthesised in plants. The fact that seven papers on ethylene were presented at the Sixth International Conference on Plant Growth Regulators held in Ottawa last summer indicates the lively interest which is now being shown in this compound.

Analysis of auxin activity

A procedure which is important to many auxin physiologists is that used for the estimation of IAA levels in plant tissues. The usual procedure is to extract the macerated tissues with a suitable solvent system, separate the acidic fraction and run this on a paper or thin-film chromatogram. This is then divided into horizontal segments of equal size which are separately extracted and the extracts are submitted to auxin bio-assay, usually a wheat or oat coleoptile extension test. By this means, a histogram is obtained in which an auxin-active region due to IAA is usually prominent and recognisable by its R_F value. The presence of inhibitors is also revealed on these chromatograms. Such diagrams are frequent in the literature and it is therefore important to examine critically their real physiological value.

Two questions have to be considered. Firstly, does the amount of IAA determined by these methods really indicate the true amount present in that tissue when it is part of the intact plant? Intarakosit examined this by macerating pea shoots and extracting portions of them at different time intervals. Her work is still to be published, but one important finding was a considerable variation in the pattern of the histograms so produced. Thus, at the R_F corresponding to IAA there was appreciable activity at the 30 minute extraction time. This fell to about one-third after 60 minutes, but after 120 minutes the level had risen appreciably and was much greater than it had been at 30 minutes. It would therefore appear that between 30 and 60 minutes considerable loss of IAA, presumably due to oxidase activity, was occurring in the extracts, but thereafter there was a release of further IAA, presumably from some conjugate. The experiments were repeated with 10^{-4}M catechol present from the stage of maceration of the tissues, catechol being an antioxidant which is known to protect IAA against destruction by IAA oxidase.

It was found that the presence of catechol led to much increased quantities of IAA in the extracts in all these experiments. It is clear, therefore, that changes in the levels of IAA take place after maceration of the tissues, and the amount found in any experiment may not reflect the level of endogenous IAA present in the fresh tissues which were extracted.

The second question is whether the finding of a high IAA level on the histogram indicates a high growth potential for the tissues concerned. Of the many amino acids available to the plant, L-tryptophan appears to be outstanding in its importance as an auxin precursor. In attempts to obtain highly active auxin fractions which might not be IAA-dependent, rich sources of auxin were sought. The rapidly growing tips of the hop, vine and runner bean were chosen for this work. These were extracted by Taylor, and the extracts were submitted to auxin bio-assay. In view of the capacity of these tissues for extension growth it was expected that high auxin activity would be revealed. In fact, in all cases and in spite of the large amounts of biological material extracted, the auxin activity as assessed in the wheat cylinder test was insignificant. In further experiments, extracts of vine tendril tissue again gave negative results. No clear evidence of gibberellin activity in the extracts was obtained by the dwarf pea test, although gibberellin activity has been reported in such tissues.[16,43] It is inconceivable that the growing tips and rapidly extending cells extracted were not being supplied with auxin; since none was found, the conclusion from these experiments would be that auxin was being rapidly used up in the reactions associated with growth. This would indeed be logical if the activity of auxins had any chemical basis. However, if tissues capable of showing extremely rapid extension growth contain little or no extractable auxin, how can physiologists recording positive IAA regions on their histograms interpret their results? Instead of the 'no extracted IAA–no growth potential' concept the true state of affairs might be the opposite. All these considerations then illustrate the difficulties involved in trying to assess and interpret the auxin status of plant tissues; indeed, it may well be that some of the published results and conclusions reached in work of this type should be re-examined. This concept has already been expounded by Mer,[44] who has presented evidence that in many instances there is little correlation between the amount of extractable auxin and the growth potential of etiolated oat mesocotyls and coleoptiles.

Growth-regulating activities of isatic and cinnamic acids

A programme of auxin research which James has just completed[45] has been concerned with isatic acids. These compounds, which only exist in the form of their salts, are produced when the isatins are treated with aqueous alkali:

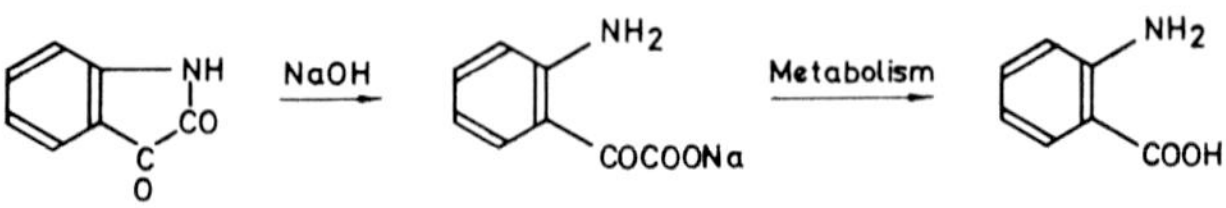

Isatin Isatic acid (Na salt) Anthranilic acid

The growth-regulating activity of isatic acid (2-aminophenylglyoxylic acid) was first shown by Haagen-Smit & Went[46] and this has been confirmed by others.[47] In the present study, the growth-regulating activity of isatic acid and 26 of its derivatives was examined.[48] In some cases, substitution with halogen led to an increase in activity; for example, 4-chloroisatic acid was more active than isatic acid in all three tests employed.[45] Metabolism experiments carried out in wheat and pea tissues with isatic and 5-chloroisatic acids revealed in each case the formation of the corresponding anthranilic acid, together with an unidentified metabolite. In order to investigate whether the activity of the isatic acid might be related to this breakdown, the whole series of the 27 corresponding anthranilic acids were synthesised and their growth-regulating activity was determined. It was concluded from these results that the observed activity of isatic acids was not due to the production of the active anthranilic acid within the tissues of the test material and that the isatic acids possess growth-regulating activity *per se*.[45,48]

It has long been known that steric configuration can be of great importance in relation to plant growth-regulating activity. With stereoisomers, only one enantiomorph is usually active.[49] Similarly, with compounds showing geometrical isomerism, activity is often related specifically to one form only. Cinnamic acid ($C_6H_5CH=CHCOOH$), for example, exists in *cis*- and *trans*- forms. That *cis*-cinnamic acid is active as a growth substance was first demonstrated by Haagen-Smit & Went.[46] In all derivatives of cinnamic acid which have since been examined, it has been found that where the *cis*- acid is active, the *trans*- isomer is not. When a cinnamic acid is synthesised, however, the product is the *trans*- isomer. Although this becomes slowly converted to the *cis*- derivative on exposure to u.v. light, a quantitative conversion has been difficult to achieve.

Using a specially designed irradiation apparatus[50] which provided for highly efficient exposure of the chemical, Ugochukwu was able to obtain good yields of the *cis*- compounds. These were then purified by fractional crystallisation and other procedures. This technique enabled a systematic study to be made of the effects of ring substitution on the plant growth-regulating activities of *trans*- and *cis*- cinnamic acids. By three different methods of assessment, most of the *cis*- acids were shown to be active and the *trans*- acids to be inactive. Substitution of fluorine, chlorine or bromine led to increased activity in the *cis*- cinnamic acids, some of these compounds being extremely active[50,51] (Fig. 2).

As with the phenoxyacetic and benzoic acids, 3,5-substitution with chlorine led to a complete removal of activity.[50] Both *trans*- and *cis*- cinnamic acids possess structural features usually considered necessary for plant growth-regulating activity, namely an unsaturated ring system and a carboxyl grouping. They differ, however, in the position of the carboxyl group in relation to the ring, for whereas in *trans*- cinnamic acid this grouping is co-planar with the ring, in the *cis*- isomer it is twisted out of the plane of the ring system. Cinnamic acid in this respect resembles other growth-regulating compounds, e.g. benzoic and 1-naphthoic acids, for which there is evidence that, for activity to be shown, the carboxyl

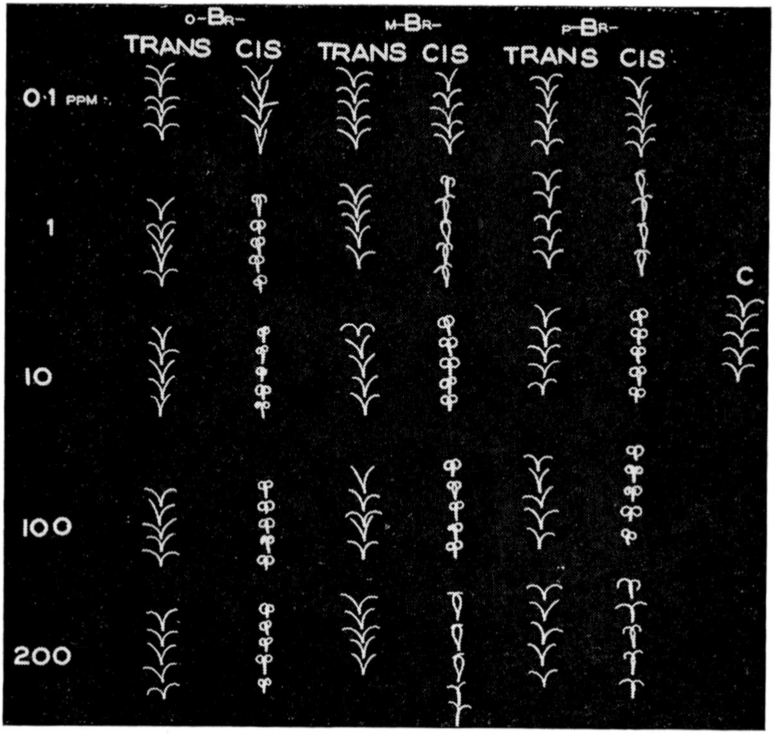

FIG. 2. *Activity in pea curvature test of* trans- *and* cis- *forms of* o-, m- *and* p- *bromocinnamic acids*
Control segments on right

group must be outside the plane of the ring.[52,53] In work at Wye College,[51] the steric and electronic effects of halogen substituents in relation to the position occupied by the carboxyl group and ease of ring hydroxylation were all considered, together with other factors which appear to be of importance in relation to the mode of action of the *cis*- cinnamic acids.

Effects of auxins on enzyme activities

It would seem logical to expect that physiologically active compounds which are capable of controlling plant growth and development must exert influences on enzyme systems within the plant. Changes in chemical composition induced by auxin treatment[54,55] are consistent with this. Furthermore, concentrations of GA as low as 10^{-10}M have been shown to initiate starch hydrolysis in barley endosperm.[56,57] It is now known that it is gibberellin secreted by the embryo of barley, wheat, oats and rice that promotes the cells of the aleurone layer to secrete the hydrolysing enzyme β-amylase. The activities of other enzymes produced by aleurone layers are also influenced by gibberellins.[58]

The effect of auxins on enzyme activity in inulin-containing plant tubers has recently been investigated.[59] Discs cut from the roots of chicory and Jerusalem artichoke were held in the surface of auxin solutions for varying periods at 25°. Considerable increases in the size and weight of the discs occurred (see Fig. 3). The effect was shown by a wide range of substances which were active in standard tests for plant growth-regulating activity. Furthermore, structural requirements which have been established for auxin activity were found to operate in the water uptake response.[60]

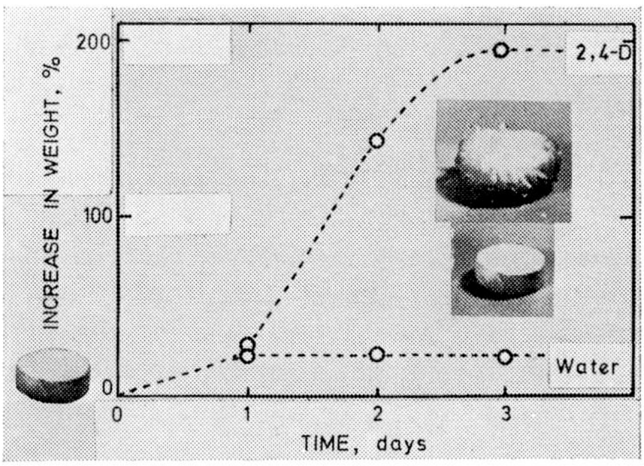

FIG. 3. *Water uptake by discs of chicory root standing in water and in a solution of* 10^{-5}M *2,4-D*

In further studies it was shown that treatment of discs of Jerusalem artichoke with 2,4-D led to a breakdown of fructosans and the production of soluble reducing sugars within the tissues and also to an increase in the rate of respiration.[59] It would therefore seem that the resulting changes of osmotic pressure and the provision of energy arising from the liberated sugars together promote the observed water uptake. Since the changes in carbohydrate status within the tissues are initiated by the growth substance treatment, Flood, Rutherford & Weston made studies on the enzyme systems which might be involved.[61] Hydrolytic enzymes were extracted from discs of chicory and Jerusalem artichoke which had been treated with 2,4-D and also from the

Table I

Water uptake and invertase and hydrolase activities of chicory and Jerusalem artichoke discs after 3 days' treatment at 25°c

Tissue	Treatment	Water uptake, increase as % of initial wt.	Units* of invertase activity x 10^6	Units† of hydrolase activity x 10^6
Chicory	Fresh tissue	—	46·3	95·9
	Water	35·2	52·5	130·5
	10^{-5}M 2,4-D	280	3392	131·5
Jerusalem artichoke	Fresh tissue	—	0	436
	Water	24	0	351
	10^{-5}M 2,4-D	96	324	1011

*1 Unit represents 2μmole hexose liberated per min at 25° per mg initial dry weight
†1 Unit represents 1μmole hexose liberated per min at 25° per mg initial dry weight

controls, and the extracts were fractionated on cellulose columns. The invertase and hydrolase activities are shown in Table I.

The increases in invertase activity observed in both tissues following treatment with 2,4-D and the enhanced hydrolase activity found in artichoke could not be associated with an increase in enzyme protein, but this finding does not rule out the possibility that a small proportion of new enzyme with very high specific activity is produced. It is also possible that the growth-substance treatment caused the enzymes to become released from a bound form on the cell wall;[62] alternatively, the activity of enzymes already present could have become greatly increased by the action of the growth substance.

Growth-regulating activity of 2,6-disubstituted phenols

Research which was developed from earlier studies on molecular compound formation[63] led in 1959 to the discovery by Wain of two highly active contact herbicides, 3,5-dibromo- and 3,5-diiodo-4-hydroxybenzonitriles (bromoxynil and ioxynil):[64]

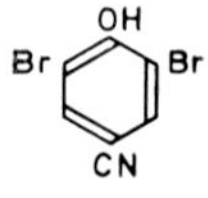

Bromoxynil

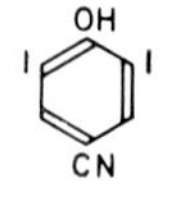

Ioxynil

Further work with Taylor on the behaviour of these compounds in the plant and in soil, and the study of analogues, led to the finding that 2,6-diiodophenol exhibits high growth-regulating activity in the tomato-epinasty, pea-segment and pea-curvature tests. Activity was also found in the corresponding 2,6-dichloro- and dibromophenols though 2,6-difluorophenol was inactive[65] (Fig. 4). More recent work with Harper has established that 2,6-disubstituted phenols with other electron-attracting groups such as $-NO_2$, $-CN$, $-CF_3$ in an *ortho-* position can also be highly active,[66] indeed, in the pea-curvature and tomato tests 2-chloro-, 2-bromo- and 2-iodo-6-nitrophenols possess activities of the same order as that of 2,4-D (Table II).

Of the dichlorophenols, only the 2,6- compound was active. The 2,3,6-trichloro- and tribromo- derivatives showed significant activity as did 2,6-dichlorothiophenol, but an important structural requirement for all active phenols is that the *para*-position in the ring must be unsubstituted. The pattern of activity shown in this new group of phenolic growth-regulators is not explicable in terms of an inhibitory action on IAA oxidase within the tissues of the

FIG. 4. *Epinastic responses in tomato plant following the application of* 40 *ml* 500 *ppm solution of the sodium salt of* 2,6-*dibromophenol to the soil*
Control plant on left

Table II

Growth-regulating activity of 2,6-disubstituted phenols in the pea-segment, pea-curvature and tomato-leaf epinasty tests

Phenol	Test						
	Pea segment			Pea curvature			Tomato epinasty
	10^{-6}M	10^{-5}M	10^{-4}M	10^{-6}M	10^{-5}M	10^{-4}M	1% applications in lanolin
2,6-difluoro-	100	100	102	0	0	0	0
2,6-dichloro-	101	102	113*	0	1	6	+
2,6-dibromo-	102	101	115*	0	2	6	+++
2,6-diiodo-	101	118*	131*	0	6	6	+++
2-fluoro-6-nitro-	100	100	100	0	0	0	0
2-chloro-6-nitro-	102	133*	124*	4	6	5T	+++
2-bromo-6-nitro-	109	133*	119*	5	6	5T	+++
2-iodo-6-nitro-	100	128*	121*	4	6	3	+++
2-fluoro-6-cyano-	100	100	100	0	0	0	0
2-chloro-6-cyano-	101	101	113*	0	1	6	+
2-bromo-6-cyano-	100	101	123*	0	4	6	++
2-iodo-6-cyano-	100	105	121*	0	5	6	+++
2,4-D	119*	135*	131*	4	6	6	+++

Pea segment results are expressed as the percentage of the final length of control segments.

Pea curvature responses are assessed on an arbitrary scale of 0 (inactive) to 6 (highly active).

Tomato epinasty tests are assessed on an arbitrary scale of 0 (inactive) to +++(highly active).

T signifies that toxic effects were observed.

*Results significantly different from control at 0·1% level.

plant, since most halogen-substituted phenols are completely inactive.

Active 2,6-disubstituted phenols do not operate as herbicides when applied to plants. They can produce severe epinastic responses on tomatoes but, except at higher dosage rates, the plants are not destroyed, and in some cases, as with 2,6-dichlorophenol, the effects disappear within a few days. Such observations indicate that these molecules are inactivated by metabolism within the plant. It may also be noted that 2-chloro-, 2-bromo-, and 2-iodo-6-nitrophenols, all of which are highly active in the pea tests, are inactive as growth-regulators in the wheat cylinder test. An explanation of such behaviour has arisen from the results of metabolism experiments which have been carried out by Harper with 2-bromo-6-nitrophenol in wheat coleoptile and pea stem tissues. Using chromatographic methods he has shown that the compound becomes rapidly converted in wheat tissues to 2-bromo-6-nitro-1,4-hydroquinone, a substance possessing no plant growth-regulating activity. Such inactivation, however, did not occur readily in pea stem tissue, which accounts, at least in part, for the activity of this phenol in the pea tests.

In both phenols and benzoic acids, high growth-regulating activity is associated with 2,6-disubstitution; also, 4-substitution destroys activity completely. Both the phenols and benzoic acids contain an acidic grouping directly attached at the 1-position of the aromatic ring and in both cases appropriate substituents in the 2,6- or 2,3,6- positions can confer growth-regulating activity. The important difference appears to be in the facility with which the phenolic growth-regulators are rendered inactive within the plant. The free *para-* position of the ring, being highly activated by the phenolic grouping in position 1-, is very susceptible to chemical attack and when this occurs, the resulting *p*-substituted phenol is inactive. The rate at which conversion of the active phenol to inactive product takes place during metabolism, may vary from species to species. Such metabolic changes will not occur to any great extent, however, in benzoic acid growth-regulators, e.g. 2,3,6-trichlorobenzoic acid, where there is no activation of the *para-* position. Such molecules are therefore able to persist within the tissues of the plant to confer a sustained herbicidal effect.

Effect of substituted benzoic acids on root growth

Work on the stimulation of seedling root growth which has also developed from research on ioxynil arose from studies with Balayannis, Zaki, Intarakosit & Taylor on the translocation in plants of the corresponding carboxylic acid, 3,5-diiodo-4-hydroxybenzoic acid, which was found to induce marked elongation of roots of cress seedlings grown in the light with the roots immersed in culture solutions containing low concentrations of the compound (see, however, work by Minarick *et al.*[67]). The dichloro- and dibromo- analogues were also active and seedlings of other species were found to show the root stimulation response. In experiments with cress and with rice, it was found that whilst pronounced cell elongation occurred in the treated roots, no significant increase in cell numbers resulted from the chemical treatment.

Other workers have shown that 3-nitro-4-fluorobenzoic acid,[67] 3,5-dibromosalicylic acid,[67] and 3,5-diiodotyrosine[68] can promote seedling root growth, and these were compared with the three 4-hydroxybenzoic acids in the cress seedling test. It can be seen (Table III) that 3,5-diiodo-4-hydroxybenzoic acid treatment produced roots up to three times as long as the controls (Fig. 5).

Table III

Root growth stimulation induced by chemical treatment

Results expressed as $\frac{\textit{length of 30 treated roots}}{\textit{length of 30 control roots}} \times 100$

Compound	Concentration, M				
	5×10^{-7}	10^{-6}	5×10^{-6}	10^{-5}	5×10^{-5}
3,5-diiodo-4-hydroxy-benzoic acid	143*	175†	285‡	331‡	217‡
3,5-dibromo-4-hydroxy-benzoic acid	110	120	210‡	234‡	261‡
3,5-dichloro-4-hydroxy-benzoic acid	102	120	160*	176†	218‡
3,5-dibromo-salicylic acid	128	144*	131	108	71
3-nitro-4-fluoro-benzoic acid	93	121	129	185†	222‡
3,5-diiodotyrosine	119	125	165*	152*	121

*Significant at the 5% level.
†Significant at the 1% level.
‡Significant at the 0·1% level.

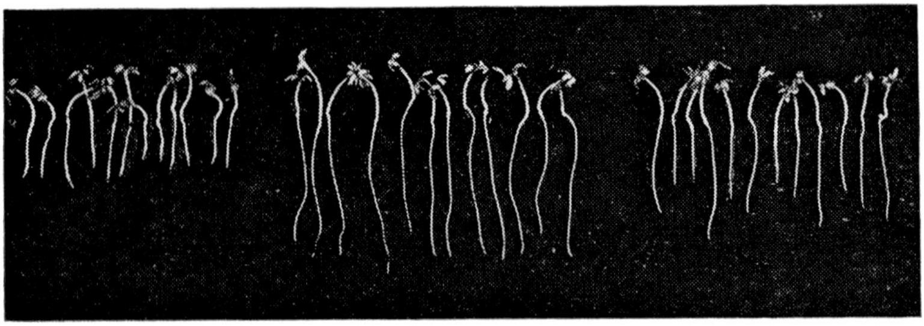

FIG. 5. *Stimulation of growth of cress seedling roots by chemical treatment*
Left to right: Roots growing in water; in 10^{-5}M 3,5-diiodo-4-hydroxybenzoic acid (sodium salt); in 10^{-6}M solution of the same compound

The corresponding dibromo- acid was also highly active; furthermore, the dichloro- derivative, whilst of lower activity, was nevertheless equal or superior to the other three compounds examined.

When cress seedlings were grown with their roots in the soil or in sand culture, the root elongation response did not occur. Since in these experiments the roots were not exposed to light, the effect of exposure was separately investigated by Taylor in water-culture experiments with and without 3,5-diiodo-4-hydroxybenzoic acid at 10^{-5}M. It was found that where the roots were held in darkness, there was no significant increase in root length resulting from the chemical treatment, although in the light the increase was obtained. It is therefore clear that the inhibition of roots by light can be removed by treatment with these substances, but other factors may also operate in this system, and these are now the subject of investigation.

References

[1] Went, F. W., *Proc. Koninke. Ned. Acad. Wetenschap.*, 1926, C**30**, 10
[2] Kögl, F., Haagen-Smit, A. J., & Erxleben, H., *Z physiol. Chem.*, 1934, **228,** 90
[3] Nenki, M., & Sieber, N., *J. prak. Chem.*, 1882, **26,** 333
[4] Ellinger, A., *Ber. dt. Chem. Ges.*, 1904, **37,** 1801
[5] Wightman, F., *Can. J. Bot.*, 1962, **40,** 689
[6] Tang, Y. W., & Bonner, J., *Arch. Biochem.*, 1947, **13,** 11
[7] Andrea, W., & Good, N. E., *Pl. Physiol., Lancaster*, 1955, **30,** 380
[8] Galston, A. W., *Bot. Rev.*, 1950, **16,** 361
[9] Van der Kerk, G. J. M., van Raalte, M. H., van Sijpesteijn, A. K., & van der Veen, R., *Nature, Lond.*, 1955, **176,** 308
[10] Rothwell, K., & Wain, R. L., *Ann. appl. Biol.*, 1963, **51,** 161
[11] Fawcett, C. H., Wain, R. L., & Wightman, F., *Nature, Lond.*, 1956, **178,** 972
[12] Taylor, H. F., & Wain, R. L., *Ann. appl. Biol.*, 1966, **57,** 301
[13] Pybus, M. B., Smith, M. S., Wain, R. L., & Wightman, F., *Ann. appl. Biol.*, 1959, **47,** 173
[14] Pybus, M. B., Wain, R. L., & Wightman, F., *Ann. appl. Biol.*, 1959, **47,** 593
[15] Kurosawa, E., *J. Nat. Hist. Soc., Formosa*, 1926, **16,** 213
[16] Adler, N., Medwick, T., Johl, R. G., & Manzelli, M. A., *Adv. Chem. Ser.*, 1961, **28,** 18
[17] Mitchell, J. W., Skaggs, D. P., & Anderson, W. P., *Science, N.Y.*, 1951, **114,** 159
[18] MacMillan, J., & Suter, P. J., *Naturwissenschaften*, 1958, **45,** 46
[19] MacMillan, J., Seaton, J. C., & Suter, P. J., *Adv. Chem. Ser.*, 1961, **28,** 18
[20] Brian, P. W., *Biol. Rev.*, 1959, **34,** 37
[21] Miller, C. O., Skoog, F., von Saltza, M. H., & Strong, F. M., *J. Am. chem. Soc.* 1955, **77,** 1392

[22] Miller, C. O., Skoog, F., Okumura, F. S., von Saltza, M. H., & Strong, F. M., *J. Am. chem. Soc.*, 1955, **77,** 2662; 1956, **78,** 1375
[23] Nitsch, J. P., *Ann. N.Y. Acad. Sci.*, 1967, **144,** 279
[24] Letham, D. S., *Life Sci.*, 1963, p. 569
[25] Letham, D. S., Shannon, J. S., & McDonald, I. R., *Proc. Chem. Soc.*, 1964, p. 230
[26] Shaw, G., & Wilson, D. V., *Proc. Chem. Soc.*, 1964, p. 231
[27] Richmond, A. E., & Lang, A., *Science, N.Y.*, 1957, **125,** 650
[28] Mothes, K., *Colloq. Int. Cent. Nat. Res. Sci.*, 1964, **123,** 131
[29] Rothwell, K., & Wright, S. T. C., *Proc. R. Soc. B.*, 1967, **167,** 202
[30] Murashige, T., & Skoog, F., *Physiologia Pl.*, 1962, **15,** 475
[31] Wright, S. T. C., *Symp. Soc. Exp. Biol.*, 1963, **17,** 18
[32] Wright, S. T. C., *Nature, Lond.*, 1961, **190,** 699
[33] Van Overbeek, J., Proc. Campbell Soup Co. Pl. Sci. Symp., 1962, p. 38
[34] Nitsch, J. P., Proc. VIth Int. Conf. Pl. Growth Regulators, 1967, in the press
[35] Wareing, P. F., Eagles, C. F., & Robinson, P. M., *Colloq. Int. Centr. Nat. Res. Sci.*, 1964, **123,** 377
[36] Addicott, F. T., Carns, H. R., Lyon, J. L., Smith, O. E., & McMeans, J. L., *Coll. Int. Cent. Nat. Res. Sci.*, 1964, **123,** 687
[37] Rothwell, K., & Wain, R. L., *Coll. Int. Cent. Nat. Res. Sci.*, 1964, **123,** 363
[38] Ohkuma, K., Addicott, F. T., Smith, O. E., & Thiessen, W. E., *Tetrahedron Lett.*, 1965, **29,** 2529
[39] Cornforth, J. W., Millborrow, B. V., & Ryback, G., *Nature, Lond.*, 1965, **206,** 715
[40] Eagles, C. F., & Wareing, P. F., *Nature, Lond.*, 1963, **199,** 874
[41] Jackson, G. A. D., & Blundell, J. B., *J. hort. Sci.*, 1963, **38,** 310
[42] Taylor, H. F., & Smith, T. A., *Nature, Lond.*, 1967, **215,** 1513
[43] Murakami, Y., *Bot. Mag., Tokyo*, 1960, **73,** 186
[44] Dattaray, P., & Mer, C. L., *Coll. Int. Centre Nat. Res. Sci.*, 1964, **123,** 475
[45] James, C. S., Thesis Univ. London, 1967
[46] Haagen-Smit, A. J., & Went, F. W., *Proc. Acad. Sci. Aust.*, 1935, **38,** 852
[47] Chen, H. R., Galston, A. W., & Milstone, F., *Pl. Physiol., Lancaster*, 1966, **41,** 1485
[48] James, C. S., & Wain, R. L., *Ann. appl. Biol.*, in the press
[49] Wain, R. L., *Sci. Prog.*, 1956, **44,** 604
[50] Ugochukwu, E. N., Thesis Univ. London, 1965
[51] Ugochukwu, E. N., & Wain, R. L., *Ann. appl. Biol.*, in the press
[52] Veldstra, H., *Ann. Rev. Pl. Physiol.*, 1953, **4,** 151
[53] Veldstra, H., 'Chemistry and mode of action of plant growth substances', 1956, p. 117 (Eds. R. L. Wain & F. Wightman) (London: Butterworths)
[54] Mitchell, J. W., *Bot. Gaz.*, 1940, **101,** 688
[55] Mitchell, J. W., Kraus, E. J., & Wighthead, M. R., *Bot. Gaz.*, 1940, **102,** 97
[56] Preleg, L. G., *Pl. Physiol., Lancaster*, 1960, **35,** 293
[57] Yomo, H., *Hakko Kyokai Shi.*, 1960, **18,** 603
[58] Varner, J. E., *Ann. N.Y. Acad. Sci.*, 1967, **144,** 219
[59] Wain, R. L., Rutherford, P. P., Weston, E. W., & Griffiths, C. M., *Nature, Lond.*, 1965, **203,** 504
[60] Rutherford, P. P., Griffiths, C. M., & Wain, R. L., *Ann. appl. Biol.*, 1966, **58,** 467
[61] Flood, A. E., Rutherford, P. P., & Weston, E. W., *Nature, Lond.*, 1967, **212,** 1049
[62] Edelman, J., & Hall, M. A., *Biochem. J.*, 1963, **88,** 36P
[63] Bennett, G. M., & Wain, R. L., *J.chem. Soc.*, 1936, p. 1108
[64] Wain, R. L., *Nature, Lond.*, 1963, **200,** 28
[65] Wain, R. L., & Taylor, H. F., *Nature, Lond.*, 1965, **207,** 167
[66] Wain, R. L., & Harper, D. B., *Nature, Lond.*, 1967, **213,** 1155
[67] Minarick, C. E., Ready, D., Norman, A. G., & Thompson, H. E., *Bot. Gaz.*, 1951, **113,** 115
[68] Mukai, N., & Kinoshita, S., *Pl. Cell Physiol.*, 1965, **6,** 161

DISCUSSION

Professor L. Fowden.—Can Professor Wain comment on the claims that tryptamine, and not tryptophan, acts as the main precursor of IAA when sterile plant tissues are employed? Is it reasonable to suggest that a symbiotic relationship between contaminating bacteria and host higher plant may be responsible for IAA production in normally growing plants?

Professor Wain.—It is extremely difficult to prove whether tryptophan is converted to IAA within the intact plant. Furthermore, this conversion may well be specific, with tryptophan as the precursor in some species and not in others. In experiments involving the incubation of excised plant tissues with tryptophan, the possibility of bacterial contamination is always present. In my view there is insufficient evidence to show that the main IAA precursor in plants is other than tryptophan. Any possibility of a symbiotic relationship between host plant and bacteria requires further investigation.

Dr. A. Winter.—As one of the workers involved in this controversy, I would like to say that under sterile conditions we have no physiological evidence indicating that tryptophan is a precursor of IAA in higher plants. In biochemical work involving extraction it is impossible to exclude contamination by bacteria and other micro-organisms; the importance of this contamination is difficult to assess. In our laboratory we have recently followed up the report of Bunt (1961) that a strain of *Nostoc muscorum* requires IAA for growth. None of 20 blue-green algae tested in our laboratory required IAA but many of them responded to IAA, stimulation of growth occurring over a range 10^{-5}–10^{-9}M-IAA. It is much easier to ensure conditions of sterility with blue-green algae cultures and we have been able to show that in certain blue-green algae, e.g. *Chloroglea fritschii*, tryptophan can act as a precursor of IAA whereas in other species it appears that tryptophan is not an IAA precursor.

CAROTENOIDS AS POSSIBLE PRECURSORS OF ABSCISIC ACID IN PLANTS

By H. F. TAYLOR
(*Agricultural Research Council, Wye College, University of London*)

Introduction

EARLY research in the field of plant growth regulation put great emphasis on the extraction and synthesis of growth stimulators. Even so the existence of growth inhibitors in plants was recognised, and by the use of filter paper chromatography both Luckwill[1] and Bennet-Clark[2] showed the presence of such an inhibitory substance in plant extracts, and the name 'inhibitor β' was proposed for it.[3] More recently it has been appreciated that the normal pattern of plant growth results from the maintenance of a balance between stimulators and inhibitors, and that both are equally important.

Inhibitors from three sources have been isolated; abscisin II[4] (from cotton bolls), dormin[5] (from sycamore leaves) and the lupin inhibitor.[6] As it was eventually shown that these all had the same chemical structure, the name abscisic acid was adopted. Abscisic acid is now known to be present in many higher plants, and is probably the major constituent of inhibitor β. This allows the early data on inhibitor β to serve as a guide to work on abscisic acid. For example it is known that plants grown normally in the light contain more inhibitor β than plants which have become etiolated through growing in the dark.[7] If abscisic acid is likewise produced by plants in response to illumination, a precursor may be found, whose conversion to this inhibitor is stimulated by light.

Such a precursor would be expected to have a wide-spread distribution in plants, like abscisic acid, and also to resemble this inhibitor in chemical structure. Such similarities are shown by certain carotenoids, and the possible conversion of these plant pigments to inhibitors on exposure to light was therefore investigated. The results of preliminary experiments devised to test this hypothesis have been published.[8]

Effect of light on pigments from Potamogeton leaves

Similarities in the ring structure of rhodoxanthin and abscisic acid, shown in Fig. 1, led initially to the investigation of this pigment as a possible precursor of inhibitory substances.

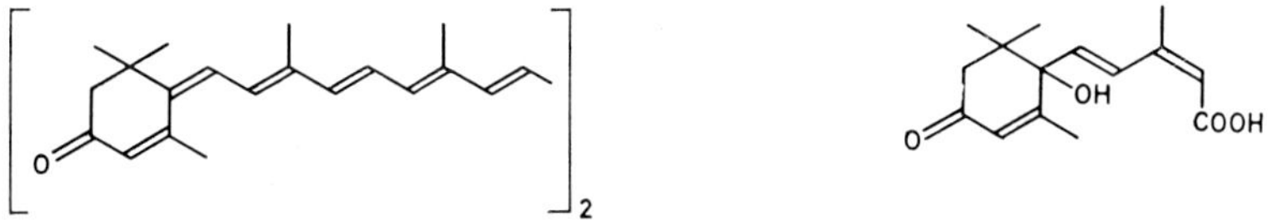

Rhodoxanthin Abscisic acid

FIG. 1.

Fresh leaves of the water-weed *Potamogeton crispus* L. were extracted with organic solvents and the pigments partially separated by thin-layer chromatography on silica gel G. using benzene:methyl alcohol (98·5:1·5) as solvent. Three main bands of carotenoids, the carotenes, rhodoxanthin and mixed xanthophylls, were removed from the chromatogram, and eluted from the adsorbent with ether. Ether solutions of the pigments were applied to discs of Whatman No. 1 filter paper in 5 cm Petri dishes and the solvent allowed to evaporate. The dishes were then kept in the dark or illuminated by a 500 watt tungsten lamp for 3 h. The light was filtered through water (5 cm path) and a stream of air directed over the dishes to reduce heating. Water (1 ml) was then used to moisten the paper in each dish and 20 cress seeds were placed on it. The dishes were then transferred to a darkened humidity box at 23·5° and after 60 h the germination of the seeds on each of the papers was assessed. For this assessment a seed was considered to have germinated when a radicle longer than 1 mm was visible. It was shown that papers which had been treated with a solution of rhodoxanthin or carotenes, whether or not they had been exposed to light, permitted normal germination. However, the germination of seeds on papers which had been treated with the mixed xanthophylls and exposed to light was negligible compared with that on similar unilluminated papers. In view of this finding, rhodoxanthin was not further investigated, and a more convenient source of xanthophyll pigments was sought.

Effect of light on carotenoid pigments from nettle leaves

Pigments extracted from the leaves of annual nettle (*Urtica dioica* L.) were saponified with alcoholic potassium hydroxide solution overnight in an atmosphere of nitrogen at room temperature. After dilution with water, this solution was re-extracted with ether to obtain the mixed carotenoids free from chlorophylls. The mixture was then partitioned between 90% aqueous methyl alcohol and petroleum ether (B.P. 60–80°) to give a hypophasic fraction which contained the xanthophylls possessing two or more hydroxyl groups and an epiphasic fraction which contained carotenes. These fractions were applied to filter paper discs as previously described and, after illumination for 1 h the papers were tested for the presence of growth inhibitors. The results of such tests (Fig. 2) showed that inhibition only occurred with the hypophasic pigments and then only after they had been illuminated, thus confirming the earlier indication that illuminated xanthophylls rather than carotenes give rise to a germination inhibitor.

The finding that not all carotenoids were equally effective led to experiments in which individual xanthophylls were isolated and tested for inhibitor production on illumination. The xanthophyll fraction of the nettle leaf carotenoids was separated on preparative layers of silica gel G. using benzene: ethyl acetate (75:25) as solvent. Better separations of the slower components were obtained when development was continued by allowing the upper edge of the chromatogram to project outside the tank. To minimise oxidation of the pigments, the main bands were removed without allowing the silica gel to become dry, and each eluted with a similar volume of solvent. Equal volumes of the solutions obtained were allowed to evaporate on filter papers and, after exposure to light or being kept in the dark, these were used in cress seed

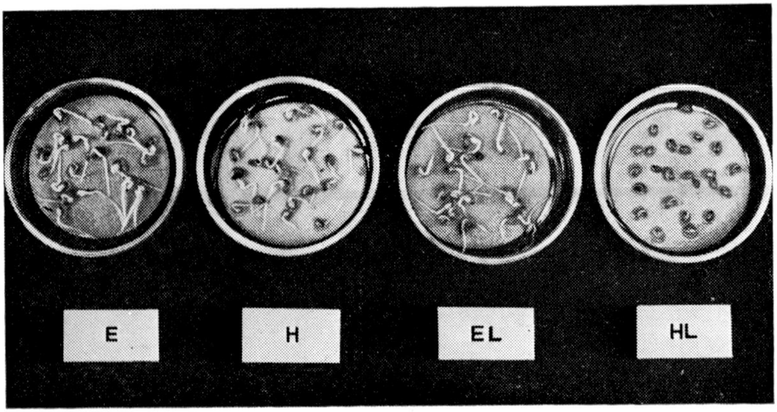

FIG. 2. *Cress seeds germinated on filter papers treated with carotenoid pigments*
Epiphasic fraction (E), after illumination (EL)
Hypophasic fraction (H), after illumination (HL);

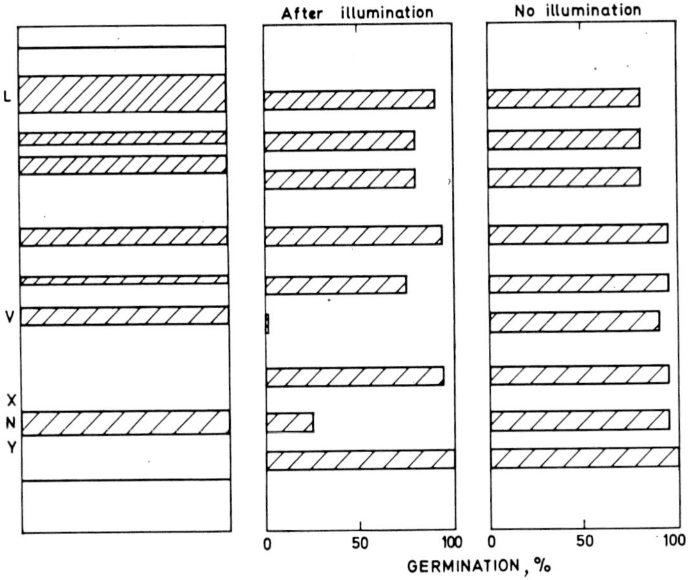

FIG. 3. *Diagram of thin layer chromatogram of nettle leaf xanthophylls* Positions shown: lutein (L), violaxanthin (V), neoxanthin (N) and two pigment-free bands (X and Y). Histograms indicate the results of germination tests carried out using the pigments before and after illumination.

germination tests. Fig. 3, which is a diagram of a silica gel chromatogram, indicates the effect on germination produced by the xanthophylls from each of the bands before and after illumination. Whilst the percentage germination values obtained with all of the pigments which had not been illuminated indicated that these xanthophylls are not themselves inhibitory, two did produce marked inhibition after exposure to light. These pigments, violaxanthin (V) and neoxanthin (N) were very effective, while much larger amounts of lutein (L) were not. Two colourless bands of silica gel (X and Y) were removed from the chromatogram adjacent to the violaxanthin and neoxanthin regions. These 'blanks' were eluted in the normal way and were shown to produce no inhibition either before or after illumination.

In further experiments carotenoid pigments occurring in leaf tissues were obtained from suitable sources, and used in equal amounts. Approximately 20 μg quantities of pigment were applied to 4·25 cm diameter filter papers and exposed to light for 1 h. The results of germination tests are shown in Fig. 4. No inhibitory

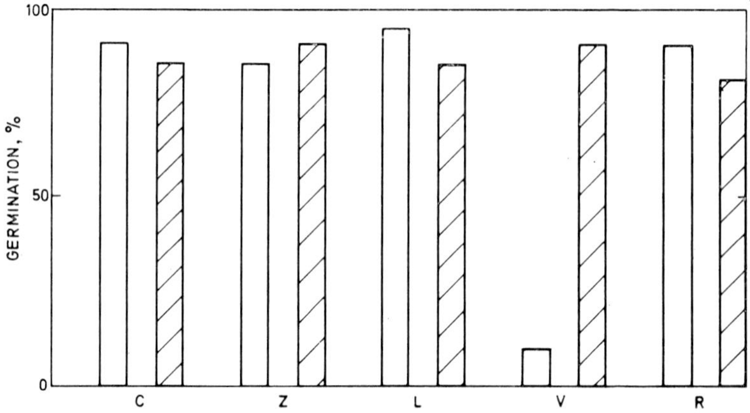

FIG. 4. *Germination tests using 20 μg carotenoid pigments*

□ illuminated ▨ unilluminated

β-carotene (C), zeaxanthin (Z), lutein (L), violaxanthin (V) and rhodoxanthin (R)

effect was found with β-carotene (from nettle leaves), lutein (from nettle leaves), zeaxanthin (from *Physalis* cups) or rhodoxanthin (from yew berries). Violaxanthin (from spinach leaves) was very inhibitory after illumination. Other tests using violaxanthin from orange peel also confirmed this finding. Thus it appears that irrespective of the source of this pigment, on illumination, a substance is formed which inhibits the germination of cress seeds.

In the introduction to this paper it was suggested that a precursor of abscisic acid might be expected to be a normal constituent of plant tissues, and that it should possess a similar structure to this inhibitor. Violaxanthin and neoxanthin must now be considered in relation to these prerequisites. The pigments satisfy the first requirement as they are amongst the most commonly occurring xanthophylls in leaf tissues.[9] They also bear an obvious structural relationship to abscisic acid as is shown in Fig. 5. Both the pigments are 5,6-epoxides of zeaxanthin, and the more effective precursor was shown to be the diepoxide. As no evidence for the production of inhibitor from zeaxanthin itself has been obtained, the possession of an 'end group' in the epoxy form may be essential. Minor modifications to this ring structure involving the breaking of an epoxy linkage, could result in the production of a hydroxyl group in the 6 position and a double bond in the ring—both modifications being present in the abscisic acid molecule. It has

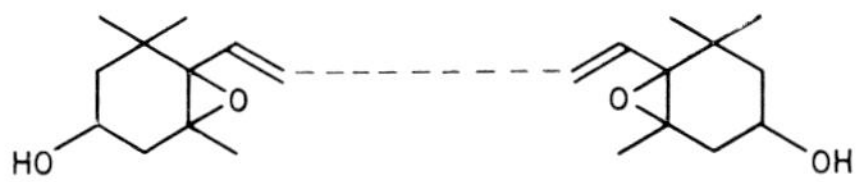

Violaxanthin

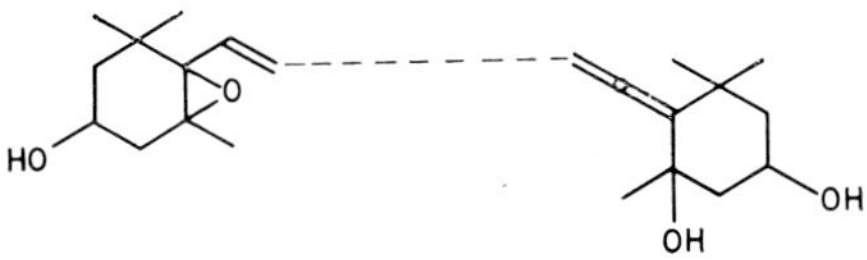

Neoxanthin

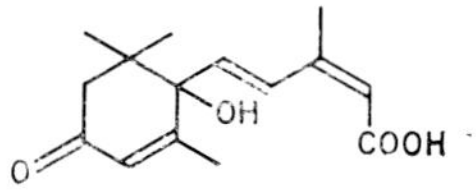

Abscisic acid

FIG. 5.

already been stated[10] that 'the functional groups of abscisin II could easily arise from a precursor like violaxanthin, a carotenoid of widespread occurrence'. The experimental results reported in the present work are in agreement with this postulate.

Production of inhibitory substances from violaxanthin

The effect of illuminating violaxanthin on filter paper for varying periods, was investigated in further experiments. Equal volumes of a solution of violaxanthin were applied to sufficient filter papers to allow two to be taken at each sampling time. When the solvent had evaporated, the papers were illuminated for periods varying from 5 min to 12 h. One paper for each time of illumination was used in a germination test, while the second was extracted with ethyl alcohol and the yellow colour assessed by measuring the optical density at 441 mμ.

The results from such an experiment are shown in Fig. 6, for which the 'percentage germination' and 'percentage of pigments remaining' are plotted against 'period of illumination'. The graphs indicate that inhibition in the cress seed germination test may be produced by illuminating the pigment for as little as 15 min. Illumination continued after 1 h appears to result in a loss of inhibitory activity as shown by the test. The pigment curve is less easily interpreted. After 2 h when the pigment level was much reduced and probably little violaxanthin remained, the germination curve began to indicate a reduction in inhibitor level. It may be deduced, however, that a period of illumination of approxi-

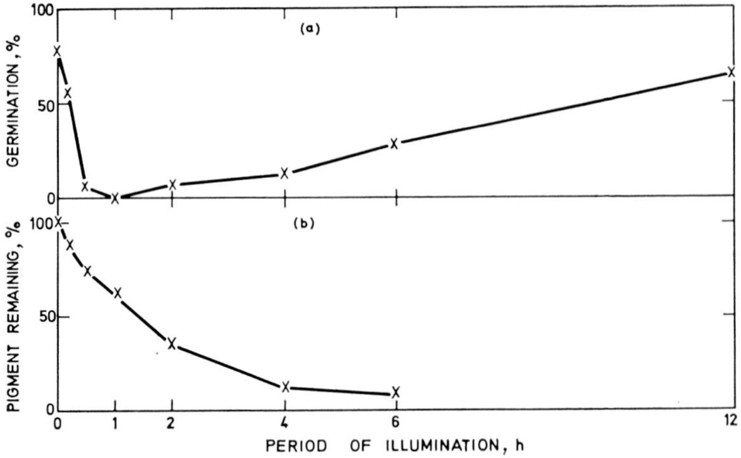

FIG. 6. *Effect of illuminating violaxanthin on filter paper for periods up to 12 h*
(a) on the subsequent germination of cress seeds
(b) on the yellow pigments which were re-extracted with ethanol

mately 1 h was optimal, under these conditions, for the production of the inhibitory effect, and this period of illumination was used in subsequent experiments.

An ether extract of filter papers on which violaxanthin had been exposed to light, was separated into acid, neutral and basic fractions. These solutions were reapplied to clean filter papers, and after evaporation of the solvent, the papers were moistened and cress seeds placed on them. Incubation for 72 h in the dark produced no germination of seeds on papers which had been treated with the neutral fraction (Fig. 7). This indicates that the inhibition had not been due to abscisic acid, but perhaps to a closely related alcohol or aldehyde which had been further converted to an acid, like abscisic acid, in the seeds.

The possibility of a light-independent conversion of the neutral inhibitor to an acidic one within the seeds was investigated. The neutral fraction of an ether extract of papers on which violaxanthin had been illuminated was obtained, and applied to a 11 cm diameter filter paper. After moistening this paper, 400 cress seeds were placed on it and incubated for 12 h in the dark. The seeds and paper were then macerated, acidified and ether extracted to give a 'treated' extract. 'Control' extracts were obtained by following the same procedure with seeds incubated for periods of 2, 6 and 12 h on moistened clean filter papers. These 'treated' and 'control' ether extracts were then partitioned into acidic and non-acidic fractions, and the acidic fractions tested for inhibition. Fig. 8 shows that, in fact, greater inhibition was obtained with the acid fraction of the treated extract than with any of the control acid fractions.

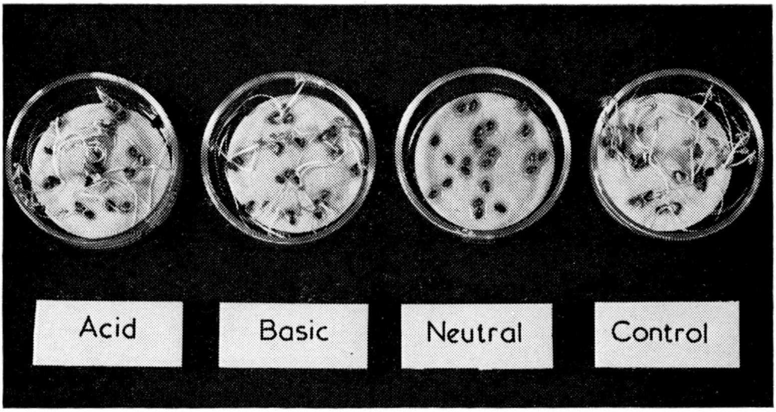

FIG. 7. *Effect of the acid, basic, and neutral fractions of an ether solution of illuminated violaxanthin on the germination of cress seed*

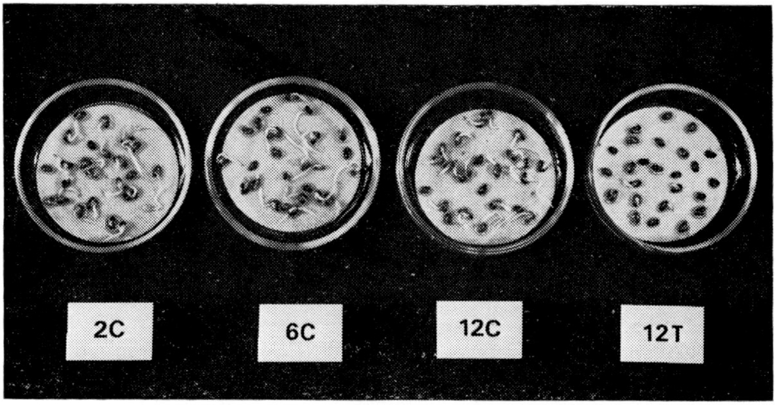

FIG. 8. *Germination tests on the acid fractions of ether extracts of seeds.*
Germination for 12 h on filter paper treated with the 'neutral' inhibitor (12T)
Germination for 2 h (2C), 6 h (6C), and 12 h (12C) on clean filter paper

This procedure was repeated in experiments in which the acid fractions of treated and control ether extracts were chromatographed on Whatman No. 20 filter paper using iso-propanol:0·88 ammonium hydroxide:water (10:1:1) as solvent. The dried chromatograms were divided into 15 equal strips and each strip assayed for inhibitor in the wheat coleoptile straight growth test.[11] The results of the assays of these chromatograms (Fig. 9) showed the presence of more inhibitor in the extract of treated seeds than in extracts of control seeds. An experiment in which more of each extract was used and synthetic abscisic acid was also chromatographed, gave the results shown in Fig. 10. The treated extract produced greater inhibition than in the previous experiment, but

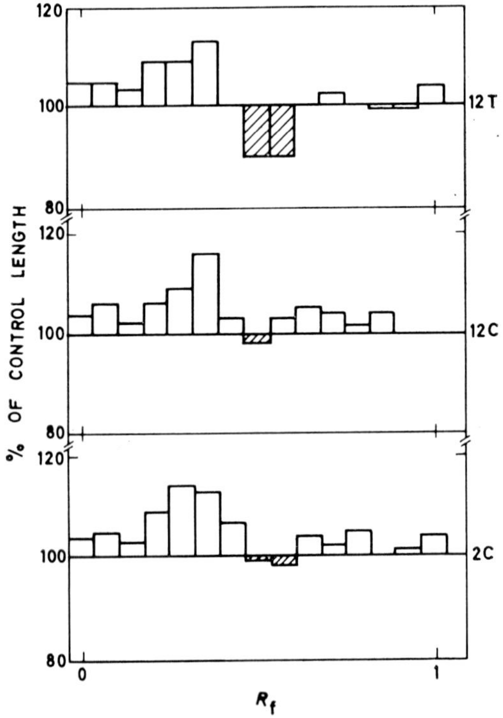

FIG. 9. *Chromatographic separation and bioassay in the wheat coleoptile straight growth test of the acid fractions of cress seed extracts of Fig.* 8

/// Histograms in the same region of the chromatogram as abscisic acid the control extract also contained a little natural inhibitor. The position of inhibitor on these chromatograms corresponded with that of synthetic abscisic acid. Thus a small but reproducible conversion of the neutral inhibitor to an acidic one, with the same R_f value as abscisic acid, is indicated.

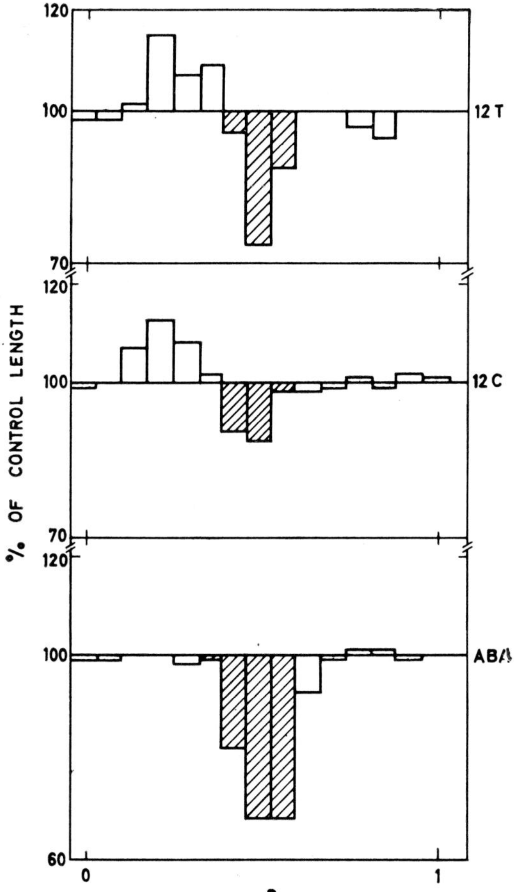

FIG. 10. *Chromatographic separation and bioassay of larger quantities of the extracts of Fig. 8*
Synthetic abscisic acid (ABA) is also included

Conclusions

The finding that a plant growth inhibitor is produced by the exposure of certain xanthophylls to light could explain some of the responses of plants to blue light. It may, for example, provide a mechanism which operates in phototropism. The initial bending of etiolated coleoptiles towards unilateral light has been explained in terms of unequal auxin distribution between the illuminated and dark sides. The photo-receptor for this effect is usually considered to be a flavin, as riboflavin is known to sensitise the photo-destruction of indolyl-3-acetic acid.[12] Voerkel, however, was the first to suggest that a carotenoid pigment might be the photo-receptor for this tropism,[13] but such an involvement was not easy to explain, especially when β-carotene was shown not to sensitise the photo-oxidation of this auxin.[14] Recently, however, Zenk[15] concluded that 'the photolysis of the auxin cannot be the cause for the induction of phototropism'. If the unilateral illumination of a coleoptile causes the production of more inhibitor on the 'light side' than on the 'dark side' then the observed curvature could be explained. This idea is supported by the report that violaxanthin is the major carotenoid in etiolated *Avena* coleoptiles[15] and also the pigment which, after illumination, gives the greatest inhibitory effect on the germination of cress seeds.

The action spectrum of the initial phototropic response of *Avena* coleoptiles has been determined in two independent investigations.[16,17] Table I shows the wavelengths of peaks in these spectra, together with the absorption maxima of violaxanthin in ethyl alcohol.

Shropshire & Withrow concluded that the active photo-receptor is a yellow pigment, probably carotenoid in nature, having absorption maxima corresponding to the peaks in the action

Table I

The location of peaks in the action spectrum of the initial photoresponse of Avena *coleoptiles and the absorption maxima of violaxanthin in ethyl alcohol*

Peaks in action spectrum, mμ		Absorption maxima of violaxanthin, mμ
After Shropshire & Withrow[16]	After Thimann & Curry[17]	After Hager & Meyer-Bertenrath[18]
470–475	474	471
440–445	445	441
410–415	425	417
370	370	—

spectrum. The data for violaxanthin indicate that this pigment satisfies these requirements, with the exception of the peak in the ultra-violet. Although *cis*-isomers of carotenoids have maxima in the long ultra-violet, they would normally be at wavelengths shorter than 370 mμ. By contrast, riboflavin has a peak in the 340 mμ region and a second peak at about 450 mμ.

From these results it is concluded that, while the function of carotenoid pigments in plants is not clearly understood, certain xanthophylls may act not only as photoreceptors but also as precursors of growth inhibitors and so have a major role in the control of plant growth by light.

Acknowledgments

The author wishes to acknowledge the major contribution made by Dr. S. T. C. Wright and Dr. T. A. Smith to this investigation,[8] and to thank Dr. M. S. Smith for the partitioning of ether extracts and Professor R. L. Wain for helpful discussions.

Thanks are also due to Dr. M. H. Zenk (Botanisches Institut der Universitat, München) for a sample of violaxanthin and to Dr. J. W. Cornforth (Shell Research Ltd., Sittingbourne, Kent), for synthetic abscisic acid.

References

[1] Luckwill, L. C., *Nature, Lond.*, 1952, **169,** 375
[2] Bennet-Clark, T. A., Tambiah, M. S., & Kefford, N. P., *Nature, Lond.*, 1952, **169,** 452
[3] Bennet-Clark, T. A., & Kefford, N. P., *Nature, Lond.*, 1953, **171,** 645
[4] Ohkuma, K., Lyon, J. L., Addicott, F. T., & Smith, O. E., *Science, N.Y.*, 1963, **142,** 1592
[5] Wareing, P. F., Eagles, C. F., & Robinson, P. M., in 'Regulateurs naturels de la croissance végétale', 1964, p. 377 (Paris: Centre National de la Recherche Scientifique)
[6] Rothwell, K., & Wain, R. L., in 'Regulateurs naturels de la croissance végétale, 1964, p. 363 (Paris: Centre National de la Recherche Scientifique)
[7] Wright, S. T. C., Thesis, Univ. Bristol, 1954
[8] Taylor, H. F., & Smith, T. A., *Nature, Lond.*, 1967, **215,** 1513
[9] Goodwin, T. W., in 'Chemistry and biochemistry of plant pigments', 1965, p. 127 (New York: Academic Press)
[10] Addicott, F. T., Ohkuma, K., Smith, O. E., & Thiessen, W. E., in 'Natural pest control agents', 1966, p. 97 (Washington: American Chemical Society)
[11] Van Steveninck, R. F. M., *Nature, Lond.*, 1958, **182,** 950
[12] Galston, A. W., *Proc. natn. Acad. Sci., U.S.A.*, 1949, **35,** 10
[13] Voerkel, S. H., *Planta*, 1933, **21,** 156
[14] Reinert, J., *Z. Bot.*, 1953, **41,** 103
[15] Zenk, M. H., *Z. Pflphysiol.*, 1967, **56,** 122
[16] Shropshire, W., & Withrow, R. B., *Pl. Physiol., Lancaster*, 1958, **33,** 360
[17] Thimann, K. V., & Curry, G. M., in 'Comparative biochemistry', 1960, Vol. I, p. 243 (New York: Academic Press)
[18] Hager, A., & Meyer-Bertenrath, T., *Planta*, 1967, **76,** 149

DISCUSSION

Dr. H. Smith.—In view of the apparent ease with which inhibitory substances can be produced from carotenoids under your conditions, could you comment on the possibility that the presence of inhibitors in other plant extracts may, in some cases, be an artefact of extraction?

Mr. Taylor.—Our results suggest that the inhibitor is not acidic, and therefore abscisic acid is not likely to be produced in this way during the extraction of plant tissues. The level of neutral inhibitor could be increased, and for this reason it would seem a reasonable precaution to carry out extraction and chromatographic procedures with the minimum of illumination.

Dr. W. A. Vincent.—In Fig. 6 the rates of destruction of violaxanthin and the appearance of inhibitor do not correspond. I know that the inhibition is not on a linear ordinate, it is more like a log-dose/response curve, but in one time region this curve is rising, i.e. there is less inhibition, whilst the pigment is still being destroyed. Can you explain this?

In addition there is a real possibility of a photo-oxidation, mediated by a carotenoid, of another compound such as a component from the T.L.C. solvent. Have you tried different solvents to test this?

Mr. Taylor.—It is true that the bleaching of yellow pigments is still occurring at a time when the inhibitory effect in germination is being reduced. One possible explanation is that inhibitor is being lost more rapidly than it is being produced. Furthermore the pigment curve does not refer specifically to violaxanthin, and therefore, after a 2 h illumination period, little violaxanthin may in fact remain.

The inhibitor is produced by the illumination of violaxanthin supplied as an alcoholic solution by Dr. Zenk. It is also produced from pigment which I have isolated from nettle leaves and orange peel employing a different method of purification to that used by Dr. Zenk. Inhibition of the germination of seeds is also produced by the illumination of mixed xanthophylls which have not been subjected to thin-layer chromatography. I think it unlikely, therefore, that a carotenoid mediated photo-oxidation of a solvent impurity is responsible for this effect.

Dr. L. J. Morris.—Bearing in mind that the light treatment of violaxanthin on filter paper is a considerably more simple model than the likely state in the intact plant cell, can Mr. Taylor comment further on the cleavage mechanism both on filter paper and in the plant and, in particular, on the role of oxygen in this? Have you carried out any experiments in which the light treatment was effected in an inert atmosphere of nitrogen or argon? If so, was any inhibitor then produced?

In addition, has any work been done on actual plant systems using radiolabelled xanthophylls as substrates, to determine directly if abscisic acid is produced? It may be that rhodoxanthin is converted enzymically although being completely inactive in the model system.

Mr. Taylor.—The illumination of mixed xanthophylls over phosphorus pentoxide or alkaline pyrogallol solution resulted in a reduction in the inhibitory effect. From this we concluded that water and oxygen were required in this conversion, but whether they are concerned in the cleavage of the polyene chain is not known.

It is true that rhodoxanthin may be converted in the plant, but not in the model system described. I now think, however, that a conversion requiring the introduction of a hydroxyl group into the 6 position is most easily achieved if the pigment is a 5,6 epoxide. It is surely less likely to occur with a 'retro' xanthophyll like rhodoxanthin.

RECENT STUDIES OF ENDOGENOUS PLANT GROWTH SUBSTANCES USING COMBINED GAS CHROMATOGRAPHY–MASS SPECTROMETRY

By J. MacMILLAN and R. J. PRYCE
(School of Chemistry, The University, Bristol)

Combined gas chromatography–mass spectrometry (g.c.–m.s.) of the gibberellins as the methyl esters and their trimethylsilyl ethers provides an unambiguous method for the detection and identification of these plant growth hormones. This is illustrated by a current and detailed study of the growth substances present in immature seed of *Phaseolus multiflorus*. Gibberellins A_1, A_5, A_6, and A_8, previously shown to be present in these seeds by isolation, have been identified in the crude acid extract, together with gibberellin A_{20} (*Pharbitis* gibberellin) and a new gibberellin A_{17} whose isolation and structure determination is briefly described. After column chromatography of the crude acid extract on a charcoal–Celite column, these gibberellins have been detected in the appropriate fractions together with gibberellin A_{19} (bamboo gibberellin). The following three non-gibberellins have also been detected and isolated: (i) two closely related compounds, G-α $C_{24}H_{35}NO_6$ and G-β $C_{23}H_{36}O_6$ which respectively retard and promote the extension of oat mesocotyl sections and which are believed to be terpenoid and (ii) phaseic acid, $C_{15}H_{20}O_5$, believed to be structurally related to abscisic acid. G.c.-m.s. studies of compound *b*, previously obtained from seed of *P. multiflorus* by Jones, supports the structure assigned by him. G.c.–m.s. has also been successfully used to detect abscisic acid in a crude extract of apple juice; the limit of detection in this case was found to be less than one part in 10^7 fresh weight of apple fruit.

Introduction

THE identification of endogenous plant growth substances by isolation is not always possible. Thin-layer chromatography,[1,2,3] combined with bio-assay and sulphuric acid induced fluorescence, has been useful in identifying known gibberellins in plant extracts, particularly when extended to include the chromatography of derivatives.[4,5,6] However, these chromatographic methods suffer from two disadvantages. First, they require authentic specimens for direct comparison of an increasing number of gibberellins which are not always available. Secondly, they can provide only circumstantial proof of identity. In seeking methods without these disadvantages attention has been turned to combined gas chromatography–mass spectrometry (g.c.–m.s.). This paper reviews progress in a current study of growth substances present in seed of *Phaseolus multiflorus* using this powerful analytical method.

Methods

As a first requirement, satisfactory methods were developed for the separation of gibberellins by gas chromatography (g.c.). These methods, an extension of those of Ikekawa *et al.*[7,8] have been described[9] and comprise the use of the methyl esters and, where applicable the trimethylsilyl ethers (TMSi ethers), each derivative being chromatographed on silanised, packed columns of 2% QF-1 and 2% SE-33 (or 30). The retention times of the 23 gibberellins, at present known, are shown in Table I and the structures of the most recently discovered gibberellins are shown in Fig. 1 together with their A numbers according to the proposals of MacMillan & Takahashi.[10]

Table I

Retention times of gibberellins A_1 to A_{23} as the methyl esters and their TMSi ethers

Gibberellins	Retention Times, min.			
	Methyl esters		Trimethylsilyl ethers of Methyl esters	
	2% QF-1(a)	2% SE-33(b)	2% QF-1(c)	2% SE-33(d)
A_1	19·7	14·7	16·3	16·8
A_2	20·7	15·1	23·1	8·2
A_3	20·6	17·6	19·1	18·3
A_4	8·4	7·4	11·2	8·9
A_5	10·4	6·7	11·3	8·1
A_6	17·1	9·4	19·1	11·4
A_7	9·1	7·9	12·8	9·5
A_8	38·6	30·7	20·7	29·3
A_9	4·3	3·9	(7·4)*	(4·5)*
A_{10}	10·4	6·9	16·0	10·8
A_{11}	6·9	5·1	(12·3)*	(6·0)*
A_{12}	2·0	4·0	(3·5)*	(4·9)*
A_{13}	6·2	11·9	6·2	12·2
A_{14}	4·8	8·4	4·5	8·7
A_{15}	14·6	9·9	(24·9)*	(12·4)*
A_{16}	16·7	15·3	17·3	14·6
A_{17}	6·5	10·1	6·4	11·5
A_{18}	10·7	16·5	6·9	13·3
A_{19}	8·7	9·9	9·8	12·9
A_{20}	9·8	6·8	10·8	8·3
A_{21}	25·1	17·0	29·3	19·3
A_{22}	18·6	11·8	17·0	18·5
A_{23}	19·7	24·5	12·9	22·1
5 α-cholestane	3·1	20·4	5·1	23·3

(a) Column temp., 201; carrier gas, N_2, 60 ml. min^{-1}
(b) Column temp., 190; carrier gas, N_2, 80 ml. min^{-1}
(c) Column temp., 179; carrier gas, N_2, 84 ml. min^{-1}
(d) Column temp., 187; carrier gas, N_2, 75 ml. min^{-1}
* Methyl ester only.

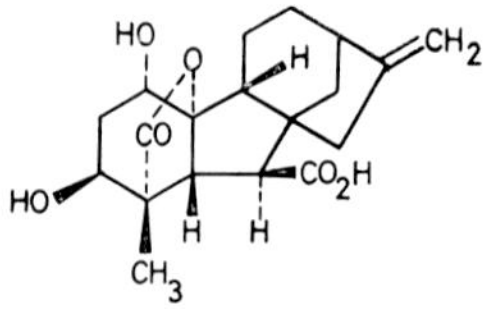

A_{16}

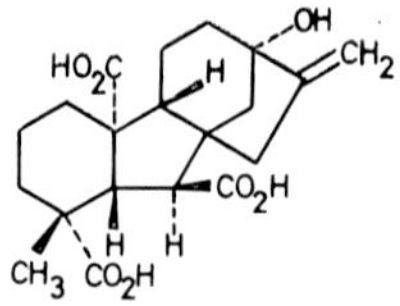

A_{17}

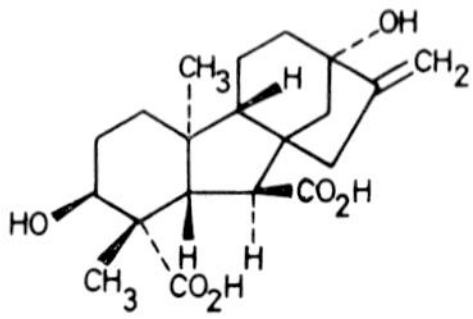

A_{18} (Lupinus - I)

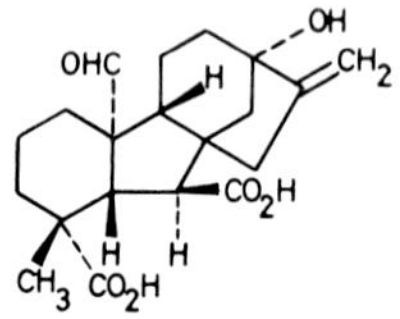

A_{19} (Bamboo)

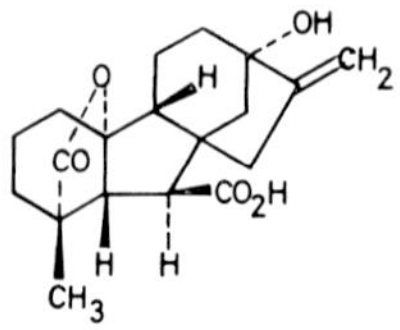

A_{20} (Pharbitis)

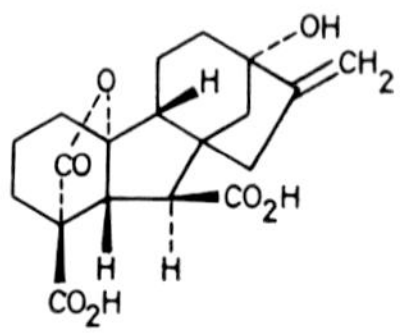

A_{21} (Canavalia - I)

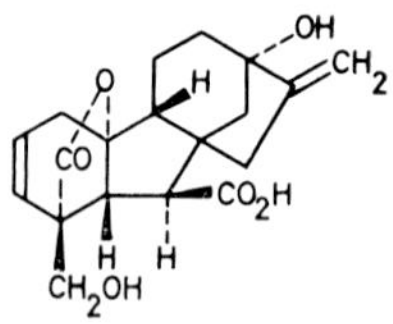

A_{22} (Canavalia - II)

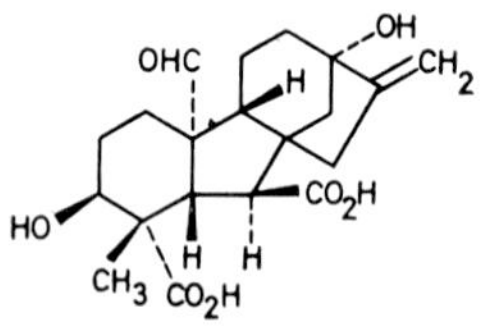

A_{23} (Lupinus - II)

FIG. 1. *Structure of recently discovered gibberellins A_{16} to A_{23}*

References

A_{16} Galt, R. H. B., *Tetrahedron*, 1968, **24,** 1337

A_{17} Pryce, R. J., & MacMillan, J., *Tetrahedron Lett.*, 1967, 4173

A_{18} Koshimizu, K., Fukui, H., Kusaki, T., Mitsui, T., & Ogawa, Y., *Tetrahedron Lett.*, 1966, 2459

A_{19} Murofushi, N., Iriuchijima, S., Takahashi, N., Tamura, S., Kato, J., Wada, Y., Watanabe, E., & Aoyama, T., *Agr. Biol. Chem., Tokyo*, 1966, 30, 917

A_{20} Takahashi, N., Murofushi, N., Yokota, T., & Tamura, S., *Tetrahedron Lett.*, 1967, 1065

A_{21} & A_{22} Takahashi, N., Murofushi, N., Yokota, T., Tamura, S., Kato, J., & Shiotani, Y., *Tetrahedron Lett.*, 1967, 4861

A_{23} Koshimizu, K., Fukui, H., Inui, M., Ogawa, Y., & Mitsui, T., *Tetrahedron Lett.*, 1968, 1143

Most of the gibberellins can be distinguished by two g.c. runs, namely the methyl esters on 2% QF-1 and the TMSi ethers of the methyl esters on 2% SE-33, although the TMSi ethers of gibberellins A_4 and A_7 methyl esters are best separated on the QF-1 column. Gibberellins A_5 and A_{20} are not well separated as either derivative on either column but this difficulty can be overcome by g.c.–m.s. as described below.

The use of the TMSi ethers is important, particularly for the methyl esters of di- and tri-hydroxylated gibberellins which have long retention times and tend to decompose during g.c. Also the use of these TMSi ethers enables the number of hydroxyl groups to be determined by subsequent g.c.–m.s.

G.c.–m.s. data were recorded with an LKB 9000 instrument using these g.c. conditions.

Results and discussion

The unfractionated acids, soluble in ethyl acetate, from the immature seed of *P. multiflorus* were first examined by g.c.–m.s. The results have been reported in a preliminary note.[11] Briefly, the crude methylated fraction, before and after trimethylsilylation, showed peaks with retention times and mass spectra identical to the authentic derivatives of gibberellins A_1, A_5, A_6, and A_8. Peaks with retention times similar to gibberellin A_{13} methyl ester and its TMSi ether gave mass spectra similar to, but distinct from, the authentic gibberellin A_{13} derivatives; this isomer of gibberellin A_{13} is discussed below. The peaks identified as gibberellin A_5 methyl ester and its TMSi ether contained an incompletely resolved component, at first thought to be the corresponding derivatives of gibberellin A_4, but subsequently shown to be those of an isomer of gibberellin A_4.

To investigate further the isomers of gibberellin A_4 and A_{13} and the unidentified components, the crude acid extract was fractionated by column chromatography on Celite–charcoal (2:1). Gradient elution with increasing percentages of acetone in water[12] gave 68 fractions.[9] The fractions were monitored by g.c. of methylated aliquots and those containing gibberellins A_1, A_5, A_6 and A_8 were readily detected.

Fraction 55, eluted with 64% acetone in water, contained the highest concentration of the isomer of gibberellin A_{13}. This new gibberellin, designated A_{17}, was isolated from fraction 55 by further chromatography on a column of Celite–silicic acid followed by thin layer chromatography on silica. The structure shown in Fig. 1 for gibberellin A_{17} was deduced from spectroscopic data which has been reported in preliminary form.[13] Briefly, the salient evidence is: (i) the methyl ester showed a parent ion in the mass spectrum with the molecular composition $C_{23}H_{32}O_7$ and was clearly a trimethyl ester from the n.m.r. spectra; (ii) the mass spectra of the TMSi ether of the methyl ester showed two features which are characteristic of a 7-hydroxylated gibberellin, namely the high intensity of the parent ion ($^M/e$ 492) and the presence of a prominent (base) peak at $^M/e$ 208; and (iii) the n.m.r. spectra of the acid and the trimethyl ester showed features characteristic of a gibberellin including a 10,10a-quartet in which the 10-proton was deshielded by the 1,4-diaxial carboxyl groups,[14] and two exocyclic methylene protons one of which shifted to lower field in going from chloroform to pyridine solution, typical[15] of the presence of a 7-hydroxyl group.

Gibberellin A_{17} is thus the 4a-carboxylic acid corresponding to the 4a-aldehyde, gibberellin A_{19} (Fig. 1), and the possible biosynthetic connection suggested that gibberellins A_{17} and A_{19} would co-occur in plants. This co-occurrence has been demonsstrated for *P. multiflorus* in the following way.

A portion of fraction 53, eluted with 62% acetone in water from the Celite–charcoal chromatogram of the crude acid fraction was methylated and examined by g.c.–m.s. The total ion current trace is shown in Fig. 2 and the peaks were scanned at the points indicated. Scan 1 is unidentified while scans 2 and 4 gave mass spectra identical to those of the methyl esters of gibberellins A_{17} and A_5 respectively. Scan 3 was identical to the published[16] and g.c.–m.s. spectrum of gibberellin A_{19} methyl ester. This result has already been reported in brief.[17]

As noted above, an isomer of gibberellin A_4 was detected in the crude acid extract, as the methyl ester and its TMSi ether as an unresolved component of the peaks of the corresponding derivatives of gibberellin A_5. This isomer has now been shown[18] in the following way to be gibberellin A_{20} (Fig. 1), first isolated from immature seed of *Pharbitis nil* by Takahashi *et al.*[19]

Although gibberellins A_5 and A_{20} are not well resolved by g.c. of their methyl esters or their TMSi ethers (Table I and Ref. 20), the mass spectra of these derivatives are, however, sufficiently different to identify these gibberellins by g.c.–m.s. of an unresolved peak. The total ion current trace of the methylated fraction 51, eluted with 60% acetone in water, from column chromatography of the crude acid fraction, is shown in Fig. 3. Although the peak is symmetrical, it has been shown to be com-

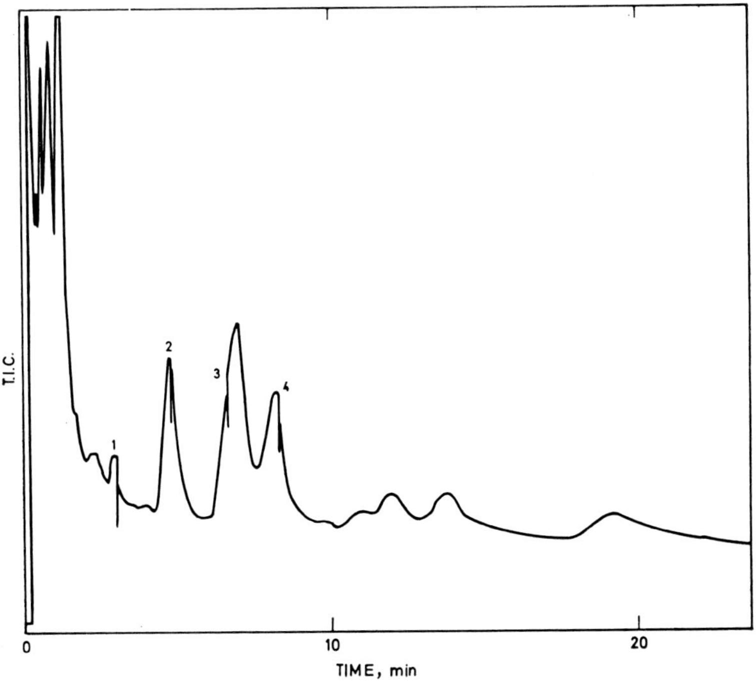

Fig. 2. *Total ion current (T.I.C.) trace of methylated fraction* 53

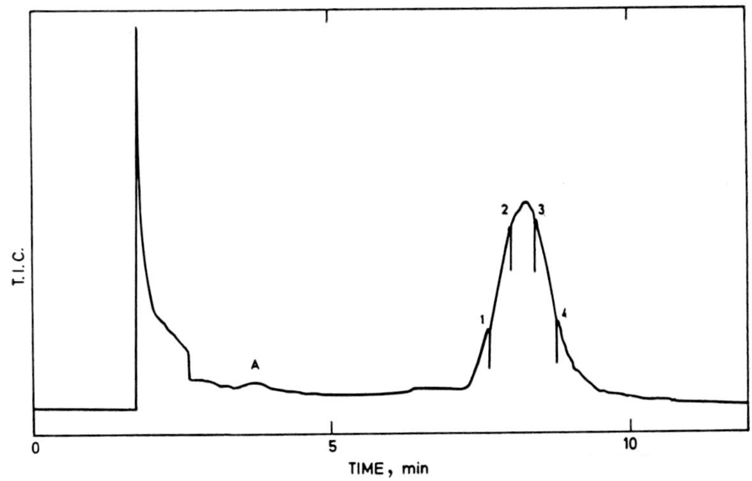

Fig. 3. *Total ion current (T.I.C.) trace of methylated fraction* 51

posed of both gibberellin A_5 and A_{20} methyl esters by scanning through the peak from $^M/e$ 0–500 in 2 seconds at the points indicated. Line diagrams of the significant peaks of scans 1 to 4 between $^M/e$ 280 and $^M/e$ 360 are compared in Fig. 4 with the corresponding portions of the spectra of authentic gibberellins A_5 and A_{20} methyl esters. The data clearly show that the scanned

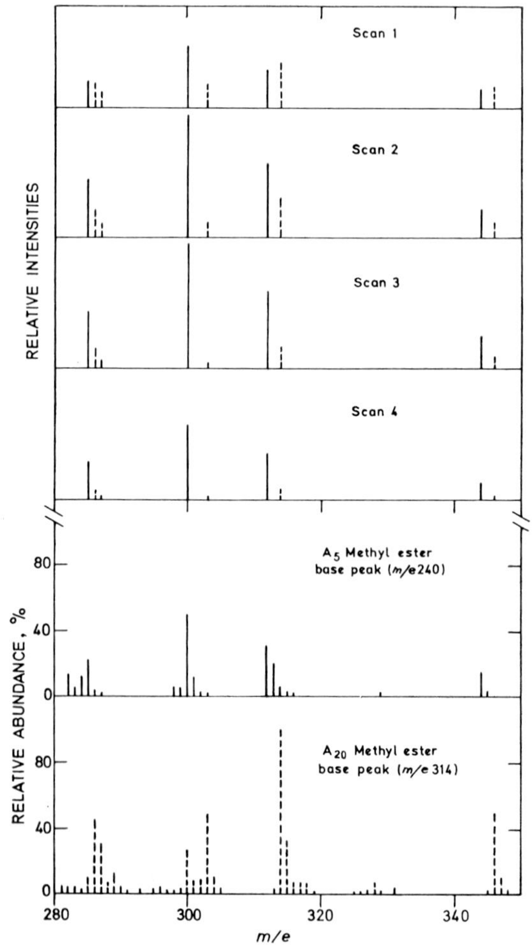

FIG. 4. *Portions of the mass spectra obtained from scans in Fig. 3 together with relevant parts of the mass spectra of gibberellins A_5 and A_{20} methyl esters*

peak is an unresolved mixture of the methyl esters of gibberellins A_5 and A_{20} in which the gibberellin A_{20} derivative precedes that of gibberellin A_5. This is more dramatically demonstrated by the method of acceleration voltage alternation (a.v.a.).[21] The a.v.a. trace of the molecular ions of gibberellin A_{20} methyl ester ($^M/e$ 346) and of gibberellin A_5 methyl ester ($^M/e$ 344) is shown in Fig. 5 and the simultaneous total ion current trace of methylated fraction 51 is shown in Fig. 6. The a.v.a. trace clearly shows the two components of which gibberellin A_{20} methyl ester has the slightly shorter retention time.

An examination has also been carried out by g.c.–m.s. on compounds *a* and *b*, obtained from immature seed of *P. multiflorus* by Jones[6] who kindly provided the remaining traces of his samples. The extant sample of compound *a* was found to be a complex mixture with no major peaks. Insufficient material was available to obtain useful mass spectra of these peaks when scanned as the methylated or methylated-trimethylsilylated derivatives. Methylated compound *b* revealed one major peak which, when scanned by g.c.–m.s., gave a mass spectrum practically identical to that of gibberellin A_6 methyl ester. Significantly, the TMSi derivative of the methyl ester showed one peak whose mass spectrum possessed a parent ion at $^M/e$ 540 with an $(M+2)^+$ peak of the correct intensity for the presence of one chlorine atom. A chlorine-containing fragment ion at M-59 was present and is typical of gibberellin A_6 methyl ester and its TMSi ether. Also the parent ion was the base peak, a strong parent ion being characteristic of the TMSi ethers of 7-hydroxy gibberellins. The above mass spectral

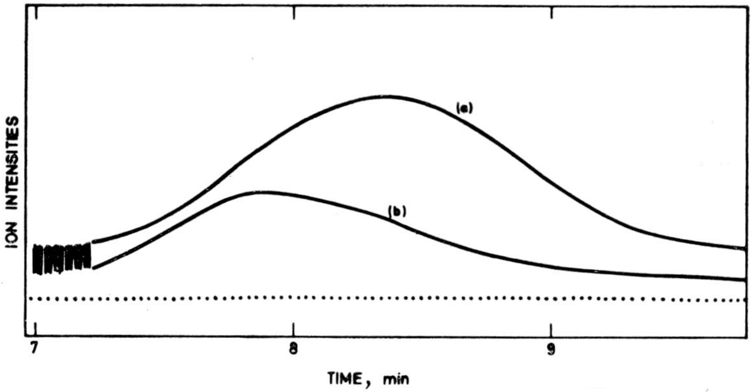

FIG. 5. *A.v.a. trace obtained from methylated fraction* 51
(a) A_5 ($^M/e$ 344) (b) A_{20} ($^M/e$ 346)

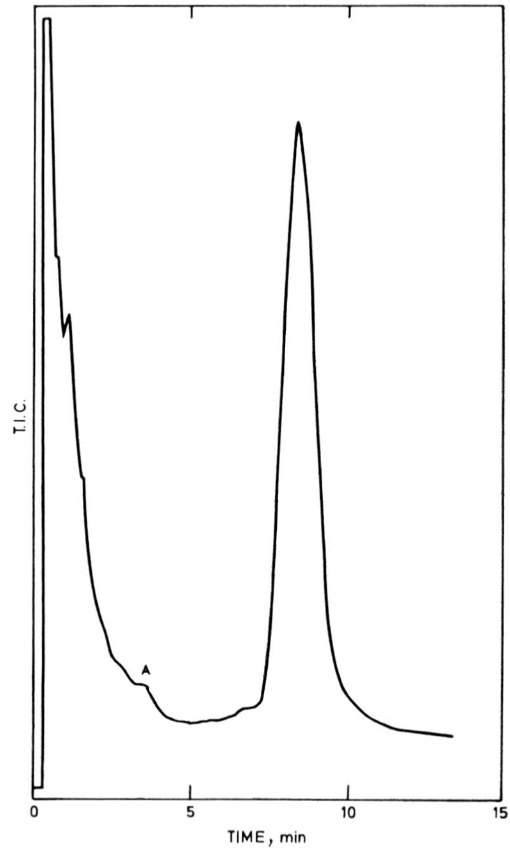

FIG. 6. *Total ion current trace of fraction* 51, *obtained during a.v.a. run*

data provide strong support for structure **I** originally suggested by Jones[6] for compound *b* and indicates that the methyl ester suffers thermal elimination of hydrogen chloride to give gibberellin A_6 methyl ester either in the molecular separator or in the ionisation chamber of the LKB 9000. The latter is more probable since it has been found that the derivative **II** of gibberellin A_6 similarly gives no parent ion containing chlorine, only the parent of the corresponding epoxide, when inserted directly into the ionisation chamber of an AEI M.S.9 mass spectrometer. In the course of the present g.c.–m.s. study of plant-growth substances in seed of *P. multiflorus* no evidence has been found for the presence of compounds *a* and *b*, thus supporting the conclusion[6] that they are artefacts.

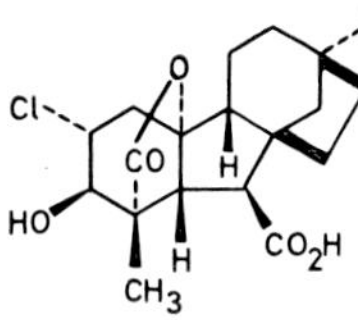

(I)

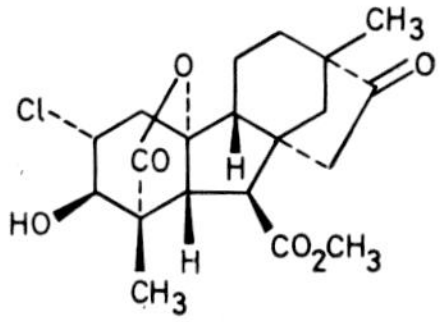

(II)

Three interesting non-gibberellin growth substances have been encountered in seed of *P. multiflorus* in the course of these g.c.–m.s. studies. Fraction 66, eluted with 86–88% acetone in water from the charcoal–Celite column of the crude acid extract, showed two main peaks by g.c. of a methylated aliquot. These two components, referred to here as G-α and G-β, became of interest when it was found that fraction 66 contained substances which promoted elongation of oat mesocotyl sections and they were isolated by column chromatography on Celite–silicic acid and by t.l.c. on silica gel. Some data are shown below for G-α and G-β which were isolated in amounts of 2·8 and 1·2 mg, respectively:

Me G-α, $^{M}/e$ 447 $\xrightarrow{-\text{HCNO}}$ $^{M}/e$ 404 $\xleftarrow{-H_2O}$ Me G-β $^{M}/e$ 422
($C_{24}H_{36}O_5$)

↑ (from G-α) ↑ (from G-β)

G-α, $C_{24}H_{35}NO_6$	G-β, $C_{23}H_{36}O_6$
1 CO_2H	1 CO_2H
No TMSi ether	bis-TMSi ether
1 >C=O	1 >C=O
$(CH_3)_2$ C<	$(CH_3)_2$ C<
1 or 2 CH_3—C≡	1 or 2 CH_3—C≡

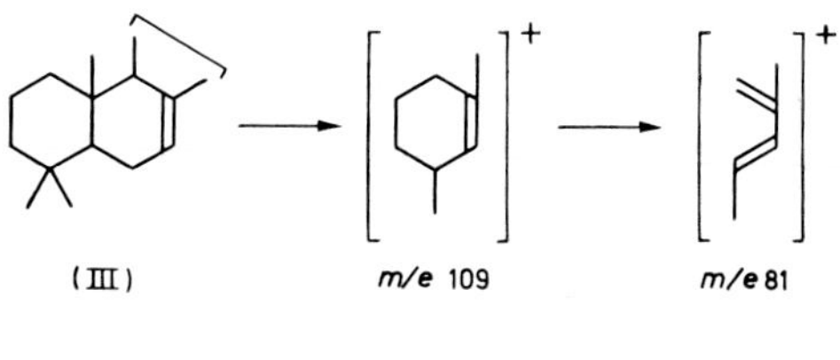

(III) *m/e* 109 *m/e* 81

Their molecular weights were determined by g.c.–m.s. and their molecular formulae of $C_{24}H_{35}NO_6$ for G-α and $C_{23}H_{36}O_6$ for G-β were established by high resolution mass spectroscopy of the methyl esters. G-α and G-β appear to be closely related since the mass spectra of the methyl esters are similar below $^{M}/e$ 404 which corresponds to the loss of HCNO from G-α and H_2O from G-β. By trimethylsilylation of the methyl ester followed by g.c.–m.s. it was shown that G-β formed a bis-TMSi ether and therefore has at least two hydroxyl groups; since G-α methyl ester did not form a TMSi ether, the two hydroxyl groups in G-β appear to be bound in some way to the HCNO fragment lost from G-α on electron impact in the mass spectrometer. Both G-α and G-β are monocarboxylic acids; both contain a second carbonyl group which may be part of an ester function and both possess three or four tertiary methyl groups, two of which seem to be geminal from the infra-red spectra. G-α and G-β appear to be terpenoid and to contain (or readily give rise to) the partial structure **III** since their mass spectra contain significant fragment ions at $^{M}/e$ 109 and $^{M}/e$ 81 characteristic[23] of such a partial structure. In one experiment G-α inhibited, and G-β promoted, oat mesocotyl extension.

The third non-gibberellin was detected by g.c. in fraction 40, eluted with 47% acetone in water from the charcoal–Celite chromatogram of the crude acid fraction from seed of *P. multiflorus*. Further chromatography of this fraction 40 on Celite–silicic acid gave gibberellin A_1 and an acid which is provisionally named phaseic acid. Phaseic acid was shown to have the molecular formula $C_{15}H_{20}O_5$ by high resolution mass spectrometry of the acid and the monomethyl ester. Spectroscopic data for the acid and methyl ester indicate the presence of the groupings shown below:

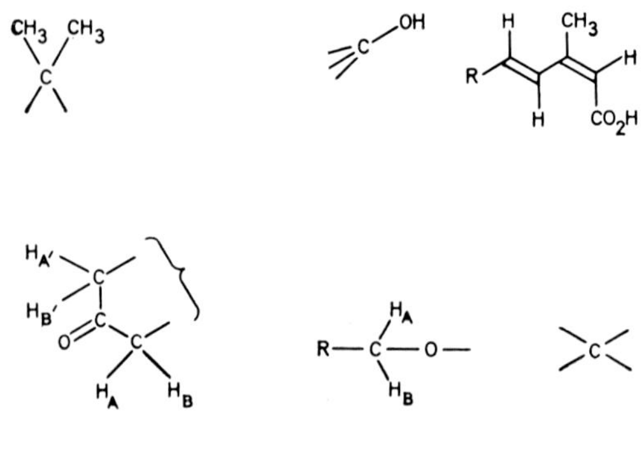

These groupings, which account for all the atoms in the molecule, can be assembled in several ways but the close similarity of the chemical shifts of the protons in the 2,4-dienoic acid in the n.m.r. spectra of phaseic acid and abscisic acid (**IV**)[22] suggest the tentative structure (**V**) for phaseic acid.

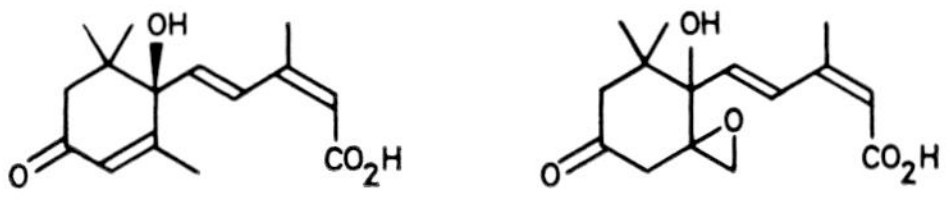

(IV) (V)

Like abscisic acid, phaseic acid does not form a TMSi ether but unlike abscisic acid, it shows only a plain and negative o.r.d. ($[\alpha]_D^{MeOH}$ −3350° (589 mμ) and −5950° (276 mμ), kindly determined by J. W. Cornforth, F.R.S.) curve. In preliminary experiments phaseic acid shows little plant growth inhibitory properties although it shows a positive but weak effect in the cotton abscission bioassay. Two facts which are perhaps not adequately explained by the tentative structure are (i) the failure to demonstrate by u.v. absorption an alkali-induced β-elimination of the epoxide oxygen to give an α,β-unsaturated ketone and (ii) the magnitude of the long-range coupling observed between one of the epoxide protons and one of the methylene carbonyl protons in the n.m.r. spectra of phaseic acid (3·0 Hz) and its methyl ester (2·0 Hz).

The use of g.c.–m.s. in the detection and identification of abscisic acid in a crude plant extract has been demonstrated recently by P. Gaskin in this laboratory. The g.c. trace of a methylated crude extract of apple juice, obtained as described by Pieniazek & Rudnicki,[24] showed two major peaks (Fig. 7); g.c.–m.s. scans of these peaks were respectively identical to scans of the corresponding peaks obtained from an aged solution of methyl (±)-abscisate (Fig. 7). Like methyl abscisate, old solutions of methyl phaseate also showed two peaks; the peaks of longer retention time are, in each case, those of the *trans, trans*-isomers of the dienoic acid side chain. G.c.-m.s. indicated that the peak of longer retention time in the crude methylated apple extract contained no methyl phaseate. The g.c. trace of the methylated apple extract shows the presence of 0·3 μg methyl abscisate in a total injection of 68 μg crude extract, corresponding to ~1 in 10^7 parts fresh weight of apple fruit.

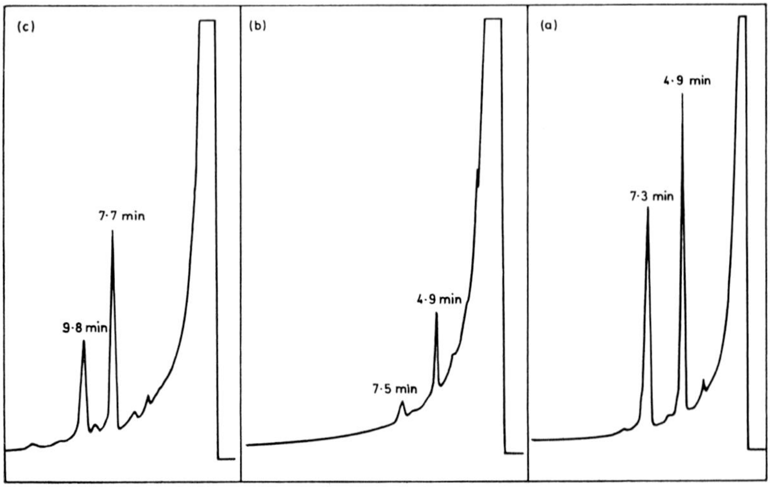

FIG. 7. *G.c. traces*
2% QF-1 on silanised 'Gaschrom Q' 183°, 60 ml N_2/min.
(a) methyl (±)-abscisate; (b) methylated crude apple extract; (c) methyl phaseate

Conclusions

The present results of this investigation by g.c.–m.s. of the growth substances in seed of *P. multiflorus* are summarised in Table II. They clearly demonstrate the power of g.c. and g.c.–m.s. in the detection and identification of endogenous plant growth substances. By the application of these techniques it has been possible to detect and identify gibberellins and other compounds which escaped notice using other methods and to show that the situation, at least in seed of *P. multiflorus*, is not simple. G.c.-m.s. has two crucial advantages. First, identification is unambiguous and, secondly, it does not require access to authentic specimens for direct comparison, provided reference mass spectra are available for comparison. To this end mass spectra by g.c.-m.s. have been obtained for the 23 gibberellins, at present known, as their methyl esters and, where applicable, as the TMSi ethers of the methyl esters. These reference spectra will be made available in a forthcoming publication and will be deposited with the Mass Spectrometry Data Centre, U.K.A.E.A., Aldermaston.

Table II

Growth Substances detected in seed of P. multiflorus *by g.c. and g.c.–m.s.*

Crude acid fraction—A_1, A_5, A_6, A_8 identified by g.c.–m.s.
A_{17} and A_{20} detected by g.c.–m.s.

↓ Celite: charcoal Column

Fraction No.	% acetone in water	Compounds
35	39–40	A_8
40	47	A_1, Phaseic acid
45	53	A_6
51	60	A_5, A_{20}
53	62	A_5, A_{17}, A_{19}, A_{20}
55	64	A_{17}
57	65–66	A_{17}
66	86–88	G-α and G-β

Concentration mg/kg seed	
A_5	1·6
A_{17}	1·8
A_{19}	0·6
A_{20}	<1·0
G-α	0·35
G-β	0·15
Phaseic acid	0·31

Acknowledgments

The g.c.–m.s. data were recorded either at the Applications Laboratory of L.K.B. Instruments, Ltd., Stockholm, or using the LKB 9000 purchased from S.R.C. grant S/BR2398 to Drs. C. J. W. Brooks and G. Eglinton (Chemistry Department, University of Glasgow). We gratefully acknowledge this co-operation and the invaluable assistance of Miss J. Malcolm and Miss H. Humphreys in recording the spectra at the University of Glasgow. We are also indebted to Professor R. Ryhage, Karolinska Institutet, Stockholm, for the use of the a.v.a. modification to the LKB 9000 instrument. Figures 3, 4, and 5 are reproduced by permission of Pergamon Press Ltd.

References

[1] Sembdner, G., Goss, R., & Schreiber, K., *Experientia*, 1962, **18**, 584
[2] MacMillan, J., & Suter, P. J., *Nature, Lond.*, 1963, **197**, 790
[3] Kagawa, T., Fukinbara, F., & Sumiki, Y., *Agric. Biol. Chem.*, 1963, **27**, 598
[4] Jones, D. F., MacMillan, J., & Radley, M., *Phytochemistry*, 1963, **2**, 307
[5] Elson, G. W., Jones, D. F., MacMillan, J., & Suter, P. J., *Phytochemistry*, 1964, **3**, 93
[6] Jones, D. F., *Nature, Lond.*, 1964, **202**, 1309
[7] Ikekawa, N., Sumiki, Y., & Takahashi, N., *Chemy Ind.*, 1963, p. 1728

[8] Ikekawa, N., Kagawa, T., & Sumiki, Y., *Proc. Japan Acad.*, 1963, **39,** 507
[9] Cavell, B. D., MacMillan, J., Pryce, R. J., & Sheppard, A. C., *Phytochemistry*, 1967, **6,** 867
[10] MacMillan, J., & Takahashi, N., *Nature, Lond.*, 1968, **217,** 170
[11] MacMillan, J., Pryce, R. J., Eglinton, G., & McCormick, A., *Tetrahedron Lett.*, 1967, p. 2241
[12] MacMillan, J., Seaton, J. C., & Suter, P. J., *Tetrahedron*, 1962, **18,** 349
[13] Pryce, R. J., & MacMillan, J., *Tetrahedron Lett.*, 1967, p. 4173
[14] Galt, R. H. B., *J. chem. Soc.*, 1965, p. 3143
[15] Hanson, J. R., *J. chem. Soc.*, 1965, p. 5036
[16] Murofushi, N., Iriuchijima, S., Takahashi, N., Tamura, S., Kato, J., Wada, Y., Watanabe, E., & Aoyama, T., *Agric. Biol. Chem.*, 1966, **30,** 917
[17] Pryce, R. J., MacMillan, J., & McCormick, A., *Tetrahedron Lett.*, 1967, p. 5009
[18] MacMillan, J., & Pryce, R. J., *Tetrahedron Lett.*, 1968, p. 1537
[19] Takahashi, N., Murofushi, N., Yokata, T., & Tamura, S., *Tetrahedron Lett.* 1967, p. 1065
[20] MacMillan, J., in 'Biochemistry and physiology of plant growth substances', in press (Runge Press, Ottawa)
[21] Sweely, C. C., Elliot, W. H., Fries, I., & Ryhage, R., *Analyt. Chem.*, 1966, **38,** 1549
[22] Ohkuma, K., Addicott, F. T., Smith, O. E., & Thiessen, N. E., *Tetrahedron, Lett.*, 1965, p. 2529
[23] Enzell, C. R., & Ryhage, R., *Arkiv Kemi*, 1965, **23,** 367
[24] Pieniazek, J., & Rudnicki, R., *Bull. Acad. pol. Sci. Cl. II Sér. Sci. biol.*, 1967, **15,** 251

DISCUSSION

Dr. G. Sembdner.—With reference to the structure of gibberellin A_{19} (bamboo-gibberellin), are the —CHO and —COOH groups at positions C-4a and C-1 respectively, free ones or does hydrogen-bonding exist between them? Also does gibberellin A_{19} behave like a dicarboxylic acid in thin-layer electrophoresis?

Dr. MacMillan.—We have not handled gibberellin A_{19}. However from the published work of Murofushi *et al.* (*Agric Biol. Chem. Tokyo*, 1966, **30,** 917.) the aldehyde group appears to be normal. I have no information on the electrophoresis of gibberellin A_{19}.

Professor P. F. Wareing.—Do you have any evidence for the natural occurrence of the *trans-trans* isomer of abscisic acid in apple juice?

Dr. MacMillan.—We have no evidence on this point.

Professor Wareing.—Will you say something about the biological activity of phaseic acid?

Dr. MacMillan.—Phaseic acid shows little or no activity in the *Rumex*, oat mesocotyl, lettuce hypocotyl or oat leaf base tests in the presence or absence of gibberellin A_3. It shows about one-fiftieth of the activity of (+)-abscisic acid in the excised wheat embryo test and about one-tenth that of (+)-abscisic acid in the cotton abscission assay.

SECOND SESSION

Chairman: Prof. R. L. Wain

PRODUCTION OF GIBBERELLIN-LIKE SUBSTANCES IN BARLEY SEED AND SEEDLINGS

by MARGARET RADLEY

(A.R.C. Unit of Plant Physiology, Imperial College, London S.W.7)

Gibberellin-like substances form in the scutellums of barley embryos during the first two days of germination. Much of the gibberellin moves out of the embryo into the endosperm and the aleurone layer. This production of gibberellin is not stimulated by mevalonic acid or sugar, but it is inhibited by (2-chloroethyl)-trimethylammonium chloride (CCC). It is possible that the scutellum produces gibberellin from a stored precursor and this possibility is discussed. Also there is some evidence that during the first few hours of germination the presence or absence of the axis affects the subsequent ability of the scutellum to secrete gibberellins. When incubated at 25°, the scutellum ceases to produce gibberellins after three days, and the axis probably commences to produce gibberellins.

Light-grown barley seedlings about one week old exude guttation drops containing trace amounts of a gibberellin-like substance although more is found in the guttation drops of wheat seedlings. The guttation drops contain virtually no activity if the plants are grown in CCC solution, which implies that the gibberellin found is being synthesised during the growth of the seedlings.

Provisional identification of the gibberellins has been carried out using paper and thin-layer chromatography and bioassays. In developing barley ears gibberellic acid (GA_3) and gibberellin A_1 (GA_1) occur accompanied during early stages of development by two unidentified gibberellins. In mature dry seed gibberellic acid is the predominant substance, but during germination the scutellum secretes a substance which is probably gibberellin A_1, with only small amounts of gibberellic acid observed. The leaf tissue of light-grown seedlings contains two substances which are probably gibberellic acid and gibberellin A_1. The nature of the active substances in guttation drops has been studied in wheat seedlings only; both gibberellins are exuded.

Introduction

It has been known for some years that gibberellic acid stimulates the synthesis of α-amylase and a number of other enzymes in germinating barley. In intact seeds gibberellic acid shortens the time taken for α-amylase to develop, while in seeds where the embryo is damaged or absent amylase will only develop if gibberellic acid is provided. Gibberellic acid is used by many maltsters to obtain quick, uniform malting of barley. It has long been known too that some substance moves from the embryo to the endosperm during germination of barley which causes the production of hydrolytic enzymes.[1] It has been suspected that this substance is a gibberellin, and most of the results described below are concerned with this.

Fig. 1 shows those parts of the barley embryo which are involved. The α-amylase which is controlled by gibberellin develops in the aleurone layer. This surrounds the starchy endosperm and is 3–4 cells thick, although where it covers the embryo it is only one cell thick. The axis is the young plant and consists of the primary root covered by the coleorhiza, and the acrospire which is the coleoptile surrounding the leaf primordia. They are connected to the node which in its turn is attached to the scutellum, a flat disc of tissue in contact with the endosperm. The scutellum is a very important part of the seed in connection with gibberellin metabolism. MacLeod & Palmer[2] have shown that the epithelial cells adjacent to the endosperm elongate and the intermediate layer between the epithelium and the endosperm breaks down. Vascular strands develop in the scutellum from the node. Lipid bodies disappear and after the first day starch granules are deposited.

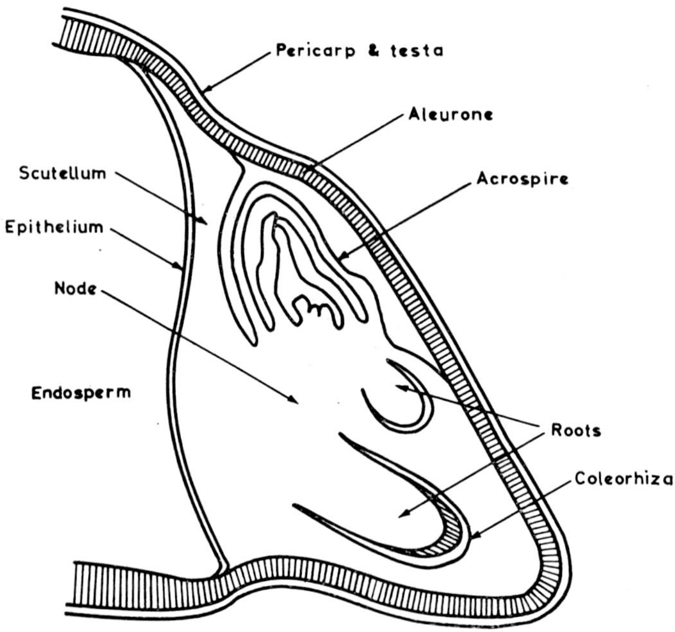

FIG. 1. *Ungerminated barley embryo*

Method used for germinating embryos

In these experiments the seed was allowed to germinate for 18–20 h, then the whole embryo was dissected. Once the intermediate layer has started to break down, this can be done without damage to the embryo. A number of embryos were placed on a small petri dish for 24 h at 25°. The agar and the tissue were frozen separately and were then extracted with ethanol. After distilling off the ethanol, the residue was partitioned between ethyl acetate and acidified aqueous solution. The residue from the ethyl acetate phase was either bioassayed directly or was first chromatographed.

The main bioassay used was the barley endosperm assay,[3] in which the half of the seed distal to the embryo is incubated with the test solution. Reducing sugars liberated by the amylase which forms diffuse into the ambient solution and are estimated using the Somogyi reagent. The quantity of sugar present is related to the amount of gibberellin in the solution.

Gibberellin production in germinating seeds

The results of incubating whole embryos or the axes and scutellums separately are shown in Table I. Considerably more gibberellin-like activity was found in the agar than in the tissues when whole embryos or scutellums were incubated, but no detectable gibberellin was found in agar after incubation of axis tissue.[4] In addition to the amounts of gibberellin, the logarithm

Table I

Amount of gibberellin-like activity in barley embryos
Dissected 18 h from beginning of germination, incubated on plain agar for 24 h, extracted from the tissues or from the agar

Sample	Gibberellin content $\mu\mu$g GA_3 equivalent per embryo		Log. total sugar	
	Axis	Scutellum	Axis	Scutellum
Tissue not incubated	13	80	2·76	3·28
Tissue (embryos incubated intact)	50	105	3·15	3·28
Agar (embryos incubated intact)		195		3·57
Tissue (axis and scutellum incubated separately)	7	85	2·60	3·30
Agar (axis and scutellum incubated separately)	0	380	2·52	3·76

Sign. Diff. log. total sugar (P=0·05) 0·28 (P=0·01) 0·37

of the sugar assayed is also shown in order to demonstrate significant differences.

If the agar contained sucrose or glucose no increase of gibberellin-like activity was found during two successive 24 h periods. When the embryos were incubated on 0·5% and 2% sucrose and were transferred to fresh dishes of agar for a second incubation period the only effect noted (Table II) was a reduction in the amount of activity secreted by the isolated scutellums during the first day on 2% sucrose. It is possible that there could be negative feed-back effect in the gibberellin–amylase–sugar system. Sugar did not stimulate gibberellin production by the axis.

Table II

Amount of gibberellin-like activity in barley embryos or scutellums
Dissected 18 h from beginning of germination, incubated on agar containing sucrose for two successive 24 h periods, extracted from the agar

Sucrose %	Gibberellin content, $\mu\mu$g GA_3 equivalent per embryo		Log. total sugar	
	Embryo	Scutellum	Embryo	Scutellum
	day 1			
0	500	290	3·86	3·67
0·5	325	115	3·71	3·36
2·0	380	65	3·77	3·17
	day 2			
0	170	300	3·49	3·69
0·5	380	225	3·76	3·58
2·0	275	185	3·65	3·52

Sign. Diff. log. total sugar (P=0·05) 0·24 (P=0·01) 0·32

(2-chloroethyl)-trimethylammonium chloride (CCC) is considered to inhibit gibberellin synthesis in both *Gibberella* cultures and higher plants. When it was incorporated into agar, and whole embryos or scutellums incubated with it, it almost entirely prevented production of gibberellin (Table III). This inhibitory effect was not reversed when the embryos were transferred to plain agar for a second incubation period.

Mevalonic acid is known to be a precursor of gibberellin synthesis in fungal metabolism,[5] and so it was incorporated in agar and intact embryos were incubated on a series of dishes of this agar for three successive days. It was necessary to chromatograph the extracts to separate the mevalonic acid from the gibberellin.

Table III

Amount of gibberellin-like activity in barley embryos or scutellums
Dissected 18 h from beginning of germination, incubated on agar containing CCC for 24 h, extracted from either tissues or agar

Sample	Gibberellin content, μμg GA_3 equivalent per embryo		Log. total sugar	
	Embryo	Scutellum	Embryo	Scutellum
Tissue not incubated	44	42	3·24	3·23
After incubation without CCC				
Tissue	190	65	3·67	3·36
Agar	465	415	3·93	3·90
After incubation on 10 μg/ml CCC				
Tissue	28	10	3·12	2·82
Agar	105	125	3·50	3·54
After incubation on 100 μg/ml CCC				
Tissue	0	9	2·69	2·78
Agar	65	38	3·15	3·20

Sig. Diff. log. total sugar (P=0·05) 0·19 (P=0·01) 0·25

Mevalonic acid had no stimulatory effect on the amount of active substances on the first two days (Table IV), but on the third day there was a significant stimulation.

When embryos were incubated for either one or three days on agar containing ^{14}C-mevalonic acid no radioactivity was detected in the gibberellin in the extracts of either the agar or the tissue, although much of the radioactivity had been incorporated into a number of substances, many of them lipids.

It seemed possible that during the first two to three days the scutellum produced gibberellin from either a bound form or from a stored precursor. It was a change that could be inhibited by CCC, but it was found that the effect of CCC was not reversed by mevalonic acid at any concentration used, and therefore the stored substance may be a later precursor in the biosynthetic pathway than mevalonic acid. A number of attempts have been made to identify this precursor but so far the results have all been negative. For example an aqueous extract of the scutellums was incubated in mildly acidic conditions for two days but no more gibberellin was obtained than had been present at the start. Another possibility was that a precursor such as (—)-kaurene could be present. As described above the scutellum is full of lipid

Table IV

Amount of gibberellin-like activity in embryos
Dissected 18 h from the beginning of germination, incubated on agar containing 10 μg/ml mevalonic acid for three successive 24 h periods extracted from the agar.

Gibberellin content, μμg GA_3 equivalent per embryo

	Plain agar		Mevalonic acid agar	
	R_f GA_3	R_f GA_1	R_f GA_3	R_f GA_1
Day 1	33	2640	52	680
Day 2	20	192	20	104
Day 3	18	25	96	96

Log. total sugar

	Plain agar		Mevalonic acid agar	
	R_f GA_3	R_f GA_1	R_f GA_3	R_f GA_1
Day 1	3·33	4·09	3·46	3·94
Day 2	3·14	3·73	3·14	3·60
Day 3	3·09	3·22	3·58	3·58

Sign. Diff. log. total sugar (P=0·05) 0·23 (P=0·01) 0·31

bodies at first so extracts of the lipids were fed back to scutellums which were no longer producing much gibberellin, but without result. It seems likely that they did not enter the cells of the scutellums. Lipid extracts from scutellums incubated with and without CCC were examined by thin layer chromatography but no differences were detected.

All the experiments referred to so far have been concerned with embryos germinated for several hours, but the first few hours of germination are also of importance. MacLeod & Palmer[2] came to the conclusion that the axis rather than the scutellum is the site of gibberellin synthesis on the basis of a different type of experiment. They incubated seeds from which different parts of the embryo had been removed and later determined the amount of α-amylase developed in the aleurone. They found that wounding the embryo in itself reduced the amylase development to some extent, but also they showed that removal of the axis 2 h after the start of imbibition, leaving the scutellum on the seed, completely prevented development of amylase, whereas if only part of the axis was left attached, especially the nodal region and root initials, some amylase development took place.

As these results did not agree with the conclusions concerning the scutellum given above, a number of experiments were carried out in which parts of the axis were removed from the dry seeds or ones which had imbibed for 2 h. After another 20 h incubation the scutellums were isolated and incubated on agar as usual. The results of one such experiment are given in Table V. There is considerable variation, but overall the removal of the whole axis has had only a slight effect whereas leaving on the nodal region is highly inhibitory. This result was obtained several times but not on every occasion and it is so far quite unexplained.

Table V

Amount of gibberellin-like activity in agar after incubation of scutellums for 24 hours

Following pretreatment of embryos 2 h after beginning of imbibition and dissection of scutellums 22 h later, three replicates

Pretreatment of embryo	Gibberellin content, $\mu\mu g$ GA_3 equivalent per embryo			Log. total sugar		
Intact	100	1900	360	3·66	3·85	3·51
Axis removed	230	230	17	3·61	3·61	3·39
Nodal region only retained	23	trace	trace	3·41	3·18	3·10

Sign. Diff. log. total sugar (P=0·05) 0·17 (P=0·01) 0·22

In order to carry out an experiment closer to those of MacLeod & Palmer half seeds with the axis treated in various ways were incubated on agar containing starch and this was then stained with iodine. The result (Fig. 2) was qualitatively like the experiment just described; removing the whole axis in this case had a marked effect, but leaving only the nodal region had a greater effect. It may be that some factor in addition to gibberellin is involved in the development of amylase, but it has been shown by Khan & Faust[6] that CCC markedly inhibits amylase synthesis in intact barley seeds and that this is reversed by added gibberellic acid, so gibberellic acid is certainly the main factor.

At the temperature used in most of these experiments, 25°, the production of gibberellin by the scutellum was much reduced by the third day, whether it was dissected from the seed after a few hours or after two days or more. When tissues were dissected

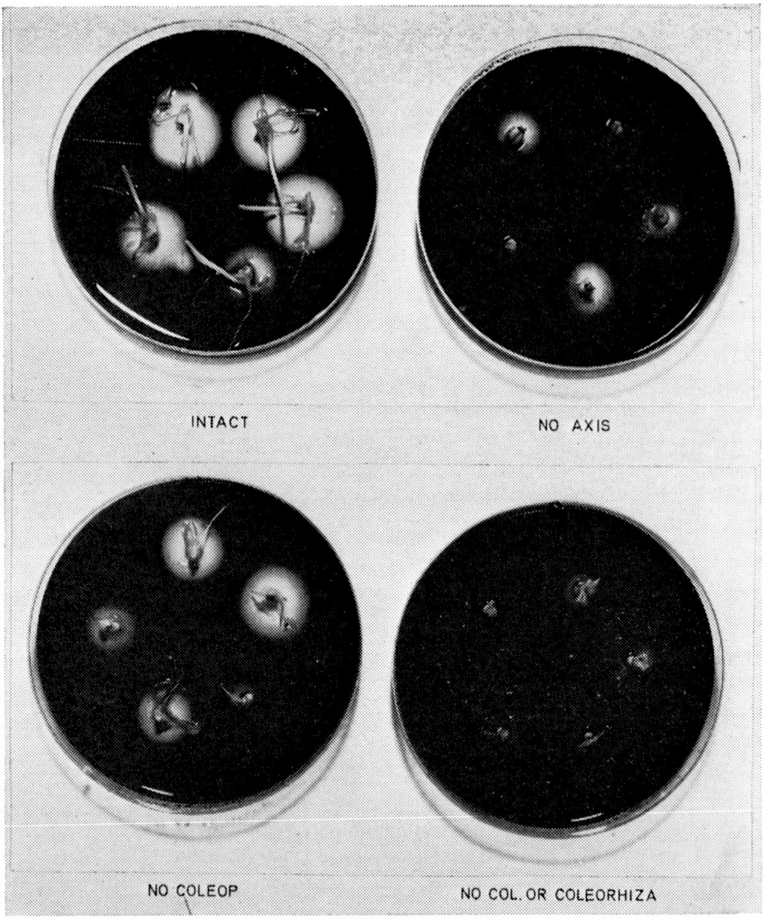

Fig. 2. *Half seeds of barley incubated on starch agar after removal of parts of axis*

54 h after the beginning of germination (Table VI) the isolated scutellum no longer produced a net increase of gibberellin. In this experiment the axis also did not produce a net increase while the intact embryo showed a small increase. These tissues were incubated on sugar agar, although no consistent response to added nutrients was obtained. In some experiments a larger increase in gibberellins produced by the axis or intact embryo was observed at this stage of growth and this agrees with the results of Yomo &

Table VI

Amount of gibberellin-like activity in barley embryos
Dissected 54 h from beginning of germination, incubated on 2% sucrose agar for 24 h, extracted either from the tissue or from the agar

Sample	Gibberellin content, $\mu\mu g$ GA_3 equivalent per embryo			Log. total sugar		
	Axis	Scutellum	Embryo	Axis	Scutellum	Embryo
Tissue not incubated	160	55	275	3·63	3·39	3·75
Tissue after incubation	42	13	115	3·31	2·95	3·55
Agar	185	50	260	3·66	3·37	3·74

Sign. Diff. log. total sugar (P=0·05) 0·17 (P=0·01) 0·24

Iinuma[7] who obtained gibberellin from barley embryos during the third to fifth days of incubation on solutions containing starch or mevalonic acid It is not clear why they failed to find gibberellin during the first two days incubation.

It seems likely that after the third day as gibberellin production by the scutellum ceases a different system starts to operate. By this time the embryo will be receiving sugar released by the endosperm. If *de novo* synthesis has not taken place up to this time it may well start to do so now. Yomo & Iinuma found that at this stage the inhibitory effect of CCC could be reversed by mevalonic acid. This suggests that CCC may act at different stages of the biosynthetic pathway in the scutellum and in the later growth of the seedling.

Gibberellin production in green plants

An investigation was then carried out using slightly older plants, week-old greenhouse grown seedlings which were photosynthesising. In addition to extracting the leaf tissue, the guttation drops were collected These are droplets which are exuded by certain cells at the tips of the leaves in conditions of high humidity. These could be collected daily without damaging the plants. They were collected by suction, and because the apparatus was always washed through with distilled water no precise measurements of volume were obtained, but as an example a close packed mass of seedlings growing in a container 12 in. × 10 in. would yield approximately 3–4 ml. The liquid was usually extracted with ethyl acetate before bioassay. Some gibberellin-like activity was regularly found in guttation drops of barley seedlings, but wheat seedlings were also used and considerably more gibberellin-like activity was found in their guttation drops The actual amount varied considerably from day to day.

When seedlings of either species were grown in water culture and CCC solution was applied through the roots, a striking reduction in gibberellin-like activity was found in the guttation drops, although the treated plants guttated quite as readily as the controls. When the plants were grown from germination on CCC solution only trace amounts of gibberellin were found in the guttation drops. If the treatment was applied after several days growth the reduction was more variable in time and scale. The results given in Table VII are for wheat plants which were 5 days old at the time of treatment. Guttation drops were collected one, two and three days after this. These results suggest that the gibberellin in the guttation drops has been synthesised in the plant since the beginning of the treatment. This method is very useful in that the plants are not damaged, but of course it is limited to those treatments which do not interfere with guttation. Treatments such as removal of part of the roots gave very variable results, probably due at least in part to variable guttation.

Table VII

Amount of gibberellin-like activity in wheat guttation drops
Taken at various times after treatment with 100 μg/ml CCC applied to roots

Days after beginning of treatment	Gibberellin content, $\mu\mu$g GA_3		Log. total sugar	
	Untreated	Treated	Untreated	Treated
1	630	96	3·19	2·68
2	930	93	3·30	2·67
3	600	249	3·16	2·92

Sign. Diff. log. total sugar (P=0·05) 0·22 (P=0·01) 0·30

Identity of the gibberellins

In the experiments described so far the results have been given as total gibberellin as estimated by the endosperm assay. Many extracts have been chromatographed either on paper or on thin layers of silica gel. They were then assayed with either the endosperm assay or with the *Rumex* leaf disc assay[8] in which the senescence of discs of dock leaves is inhibited by gibberellins.

Different gibberellins have different activities in the two assays which assists in their identification Some dose-response curves are shown in Fig. 3. The main point to notice is that gibberellin A_1 (GA_1) is at least as active as gibberellic acid (GA_3) in the endosperm assay, and possibly slightly more active in the middle of the concentration range. In the *Rumex* assay it is markedly less active. Attempts to use gas–liquid-chromatography for gibberellin identification have so far been unsuccessful.

It was shown several years ago that in developing ears of barley, GA_3 occurs accompanied by at least one other gibberellin.[9,10] GA_3 was identified by thin layer chromatography of the free acid and of various derivatives, by fluorimetry and by bioassay. The other gibberellin was very active on the cucumber seedling assay, but no fluorescence was observed on chromatograms sprayed with

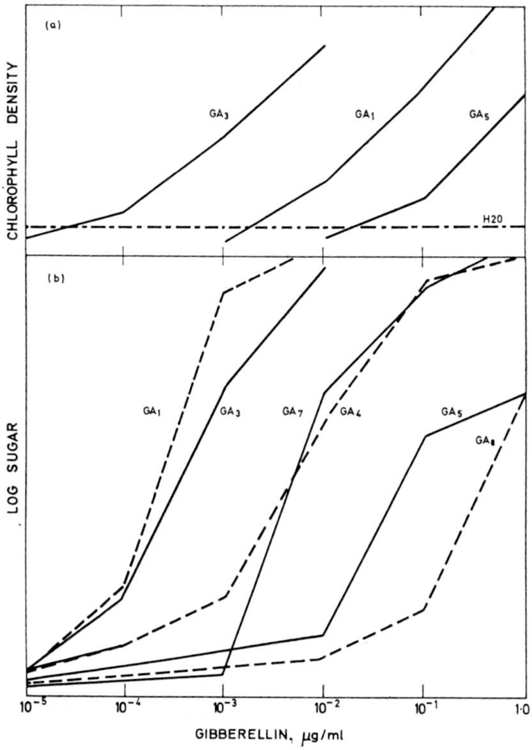

FIG. 3. *Dose-response curves of several gibberellins in two assays*
(a) *Rumex* assay (b) Endosperm assay

sulphuric acid, and it was considered to be a new gibberellin. This material was re-examined recently and it now seems probable that there are two gibberellins which have much higher R_f's than GA_3. They are active in both endosperm and *Rumex* assays, and one in particular is highly active in the cucumber seedling assay (Fig. 4). In all the histograms shown here varying amounts of

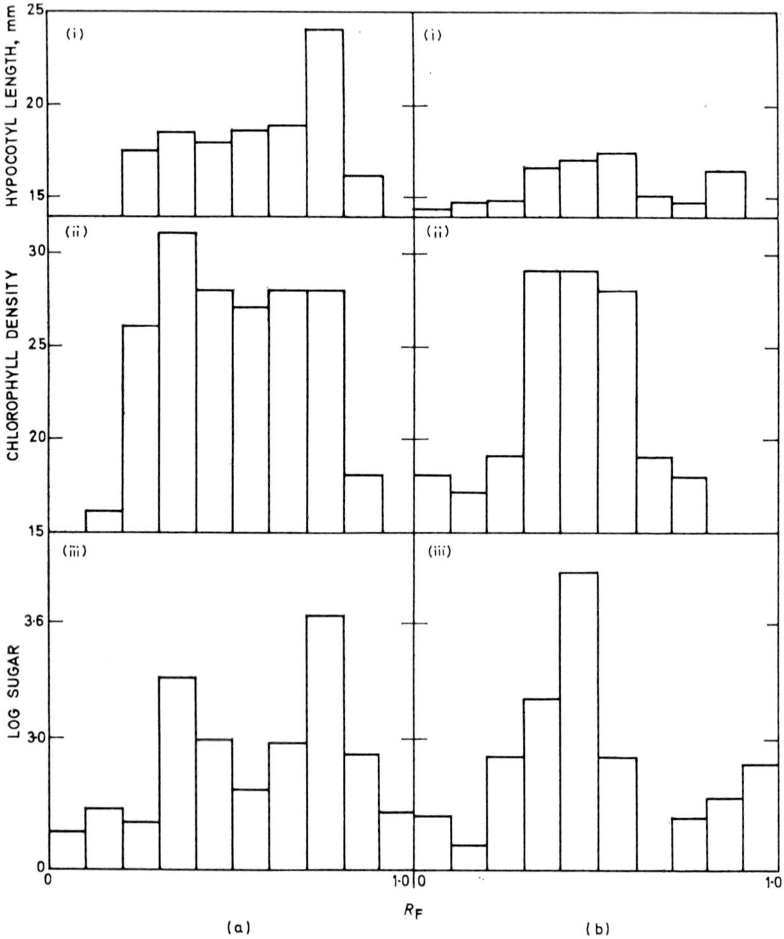

FIG. 4. *Three assays of chromatograms of immature barley extract*
Rechromatographed after elution from (a) R_f 0·4–0·8,
(b) R_f 0·2–0·4. Solvent system—benzene:acetic acid: water, 4:2:1
(i) Cucumber 20 g fresh weight
(ii) *Rumex* 130 g fresh weight
(iii) Endosperm 5 g fresh weight

extract are represented, adjusted for each assay, and they are intended for qualitative comparison only. These two gibberellins were found in barley ears harvested about ten days after flowering, but they were not detected when ears were next harvested a month later and there was no evidence of their presence in germinating seed.

The gibberellin activity that was consistently found at all stages of development, in immature seed, dry seed, germinating seed and green seedlings was always found at the R_f of GA_3 and GA_1 in all solvent systems. The two gibberellins were separated by chromatography in benzene:acetic acid:water (4:2:1)[11] for 48 h. Only that part of the chromatogram in the gibberellin A_1–A_3 region is shown in Fig. 5; no other known gibberellin has R_f values anywhere near this region. Immature ears were harvested at three stages of development. Only one batch contained detectable amounts of the two new substances described above but all three samples had peaks at positions corresponding to GA_3 and GA_1 (Figs. 5 and 6).

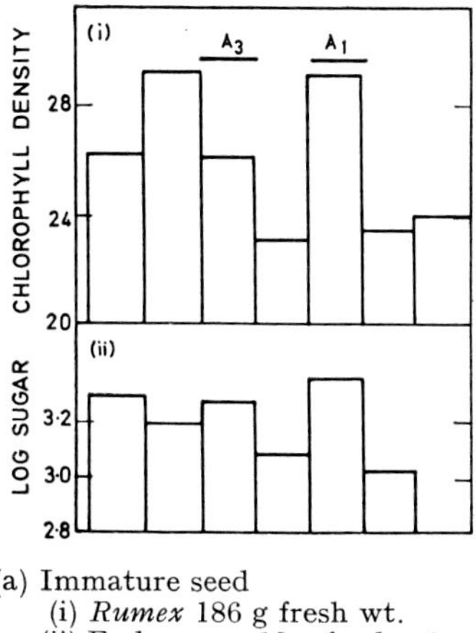

(a) Immature seed
(i) *Rumex* 186 g fresh wt.
(ii) Endosperm 10 g fresh wt.

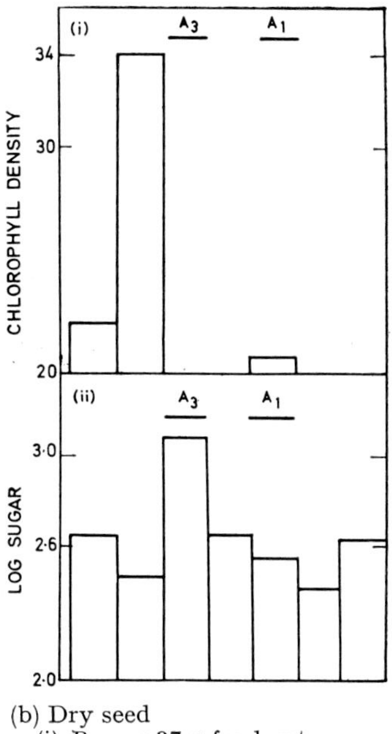

(b) Dry seed
(i) *Rumex* 37 g fresh wt.
(ii) Endosperm 10 seeds

FIG. 5. *Two assays of parts of chromatograms of various extracts*
Solvent system—benzene: acetic acid: water, 4:2:1, run for 48 h
Ordinates: for Rumex=chlorophyll density; for Endosperm=log. sugar

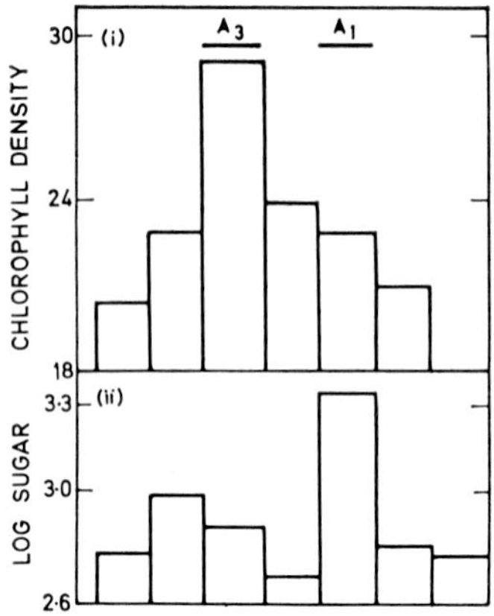

(a) Germinating seed
(i) *Rumex*. Whole seed 24 h. 24g dry wt.
(ii) Endosperm 0·25 g dry wt.

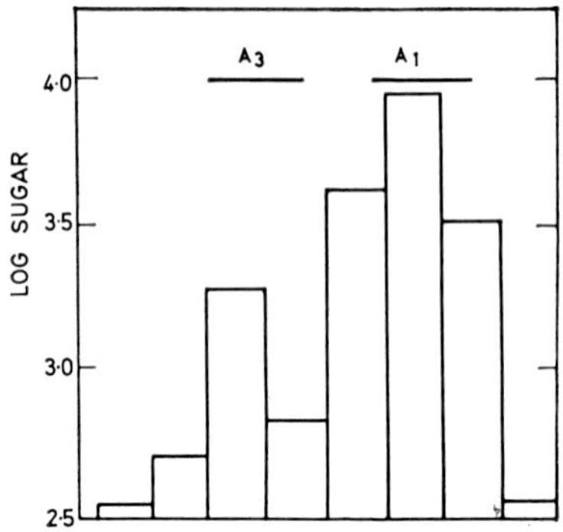

(b) Germinating seed
Endosperm 20 embryos 18 h

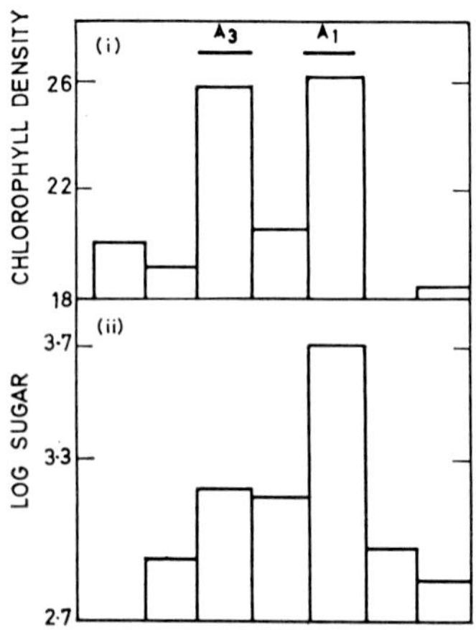

(c) Embryo exudate
(i) *Rumex* 32 embryos
(ii) Endosperm 2 embryos

FIG. 6. *Two assays of parts of chromatograms of various extracts*
Solvent system—benzene: acetic acid: water, 4:2:1, run for 48 h
Ordinates: for Rumex=chlorophyll density; for Endosperm=log. sugar

It was difficult to estimate the relative amounts of the two gibberellins but there was a marked overall decrease in concentration of activity in both zones 6 weeks after flowering. This decrease as seed approaches maturity has been found in a number of other species. The position and size of the peak of GA_3 was apparently affected by impurities in the sample, but as GA_3 has previously been identified by fluorimetry in similar extracts its identity is not in doubt. GA_1 has a much weaker fluorescence than GA_3, which would seem to be the most likely reason for the failure to observe it in previous studies.

The amount of gibberellin present in mature dry barley seed is very low. Two experiments with extracts of milled whole seed are represented in Figs. 5 and 6. GA_3 is clearly the principal substance present. As germination commences the amount of gibberellin present rises, the increase being largely, if not entirely, due to GA_1. Extracts of agar after incubation of embryos shows the same picture, mainly GA_1 but with some GA_3 present. Green leaves of week-old barley seedlings have the two gibberellins also (Fig. 7) as do green wheat leaves and wheat leaf guttation drops.

The relationship between the two gibberellins is not yet known. Pure extracts, i.e. those of diffusates into agar and of wheat guttation drops, gave GA_1/GA_3 ratios of the order of 20/1. Other extracts had different, less-reproducible ratios, due to impurities masking one gibberellin or to spread out of the peaks. Dry seed undoubtedly contains mainly GA_3 and the first production of gibberellin by the scutellum is mainly GA_1. When embryos are incubated on CCC, the GA_1 peak does not appear but there is still some activity at the GA_3 zone. It is possible that the scutellum produces only GA_1, but that both substances are formed during the rest of the life cycle of the plant. It is not yet known whether or not one gibberellin acts as a precursor of the other.

References

1 Kirsop, B. H., & Pollock, J. R. A., *J. Inst. Brew.*, 1958, **64,** 227
2 MacLeod, A. M., & Palmer, G. H., *J. Inst. Brew.*, 1966, **72,** 580
3 Nicholls, P. B., & Paleg, L. G., *Nature, Lond.*, 1963, **199,** 823
4 Radley, M., *Planta*, 1967, **75,** 164
5 Birch, A. J., Rickards, R. W., & Smith, H., *Proc. Chem. Soc.*, 1958, p. 192
6 Khan, A. A., & Faust, M. A., *Physiologia Pl.*, 1967, **20,** 673
7 Yomo, H., & Iinuma, H., *Planta*, 1966, **71,** 113
8 Whyte, P., & Luckwill, L. C., *Nature, Lond.*, 1966, **210,** 1360
9 Jones, D. F., MacMillan, J., & Radley, M., *Phytochem.*, 1963, **2,** 307
10 Radley, M., *Nature, Lond.*, 1966, **210,** 969
11 Bird, H. L., & Pugh, C. T., *Pl. Physiol., Lancaster*, 1958, **33,** 45

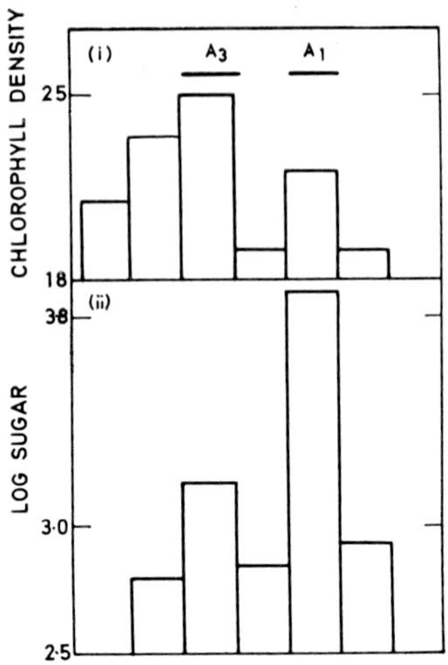

(a) Barley leaf tissue
(i) *Rumex* 40 g fresh wt.
(ii) Endosperm 3·3 g fresh wt.

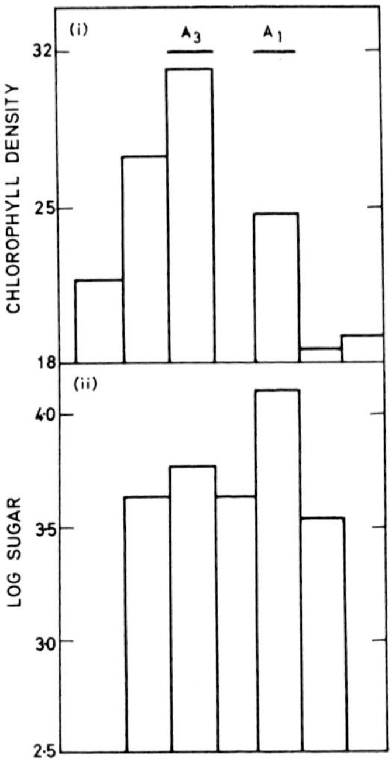

(b) Wheat leaf tissue
(i) *Rumex* 72 g fresh wt.
(ii) Endosperm 6 g fresh wt.

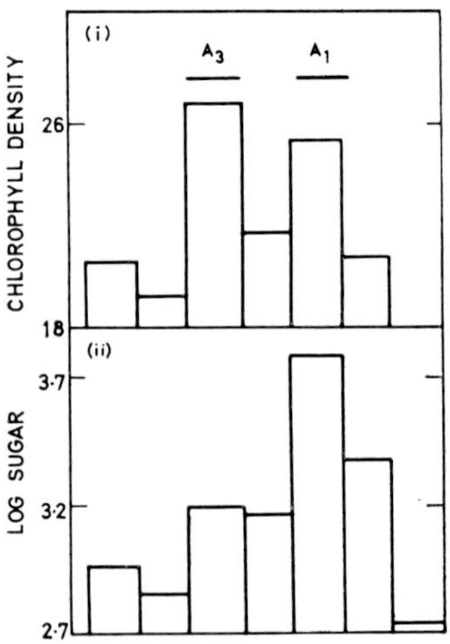

(c) Wheat guttation drops
(i) *Rumex* 4/5 total
(ii) Endosperm 1/10 total

FIG. 7. *Two assays of parts of chromatograms of extracts of wheat and barley leaves and of wheat guttation drops*
Solvent system—benzene: acetic acid: water, 4: 2: 1
Ordinates: for *Rumex*=chlorophyll density;
for Endosperm=log. sugar

DISCUSSION

Mrs. R. Verbeek-Wyndaele.—You say that mevalonic acid (MVA) has no influence on the formation of gibberellic acid (GA) in isolated embryos. We presume however that mevalonic acid does influence the formation of gibberellic acid in isolated aleurone layers, based upon the following results: 10^{-3} molar MVA and 10^{-6} molar GA induce α-amylase formation in isolated aleurone layers in the same manner and the same quantity. MVA is known to be a precursor of GA biosynthesis in *Fusarium* and other tissues, and with our results we may suggest that MVA is a precursor of GA biosynthesis in barley too.

Dr. Radley.—The difference in our results could be due to the different concentrations used in our experiments. Perhaps the influence of MVA on α-amylase formation can be ascribed to a direct effect of MVA on α-amylase formation.

Mrs. Verbeek-Wyndaele.—I think that the idea of MVA as a precursor of GA biosynthesis is justified, because at our laboratory we were able to isolate other precursors of the GA biosynthesis in barley, such as (—)-kaurene and (—)-kaurenoic acid. Therefore there is a good possibility that MVA and (—)-kaurene are involved in the biosynthesis of GA. Nevertheless the difference in our results can indeed be a consequence of different concentrations used in the experiments. In fact only one concentration of MVA, 10^{-3} molar, is able to promote the induction of α-amylase. Higher and lower concentrations give no induction at all.

Dr. Radley.—MVA is certainly a precursor of *de novo* GA biosynthesis, but in germinating embryos there may be a stored precursor from a later stage in the biosynthetic pathway.

Dr. G. H. Palmer.—I cannot agree that the scutellum deprived of its axis before 15 h, after initial wetting of the grain, is capable of gibberellin secretion. Even after 18 h we have found that it is the continued presence of the growing axis which influences the extent of α-amylase production in the endosperm, (see *Nature,* **216,** (1966) 1342). After axis stimulation the scutellum may function primarily as a transport organ, directing the gibberellin-like stimulus to the dorsal surface of the endosperm.

Dr. Radley.—I have repeatedly found that if the axis is removed 2 h after the start of imbibition and the scutellum dissected 20 h later it then secretes almost as much gibberellin-like activity as a scutellum from an intact embryo. The effect found by MacLeod & Palmer could be due to differences occurring outside the embryo, e.g. reduced translocation through the aleurone layer.

ISOLATION, STRUCTURE AND METABOLISM OF A GIBBERELLIN GLUCOSIDE

By G. SEMBDNER, J. WEILAND, O. AURICH and K. SCHREIBER

(*Institut fur Kulturpflanzenforschung, Gatersleben*)

A new polar gibberellin has been isolated in an overall yield of about $4 \times 10^{-5}\%$ from 3760 kg immature fresh fruits of *Phaseolus coccineus* L. Its structure was established as gibberellin A_8 O(3)-β-D-glucopyranoside by use of degradation studies as well as n.m.r. and other spectroscopic studies.

Investigations of the biological activity and metabolism, using isotopically labelled gibberellins, showed that this glucoside is a deactivated form of gibberellins, possibly with a depot function, which arises from more active gibberellins during seed ripening.

The occurrence of glucosides of gibberellin A_8 and gibberellin A_3 in bleeding sap of Norway maple and Wych elm, respectively, suggests that these conjugated gibberellins may be also favoured forms for translocation in plants.

Introduction

Some years ago, a number of plants were screened for their endogenous gibberellins using thin-layer chromatography and the dwarf pea bioassay. During these studies, immature seeds and pods of *Phaseolus coccineus* L. were found to contain, in addition to gibberellin A_1, A_5, A_6, and A_8 which had been already described from bean seeds, gibberellin A_3 and a much more polar gibberellin-like substance (Fig. 1).[1] This latter compound which could not be extracted from acidified aqueous solution with ethyl acetate but only with n-butanol, has been provisionally designated as '*Phaseolus* gibberellin ϵ'. Its chromatographic behaviour proved to be different from that of the other known gibberellins.

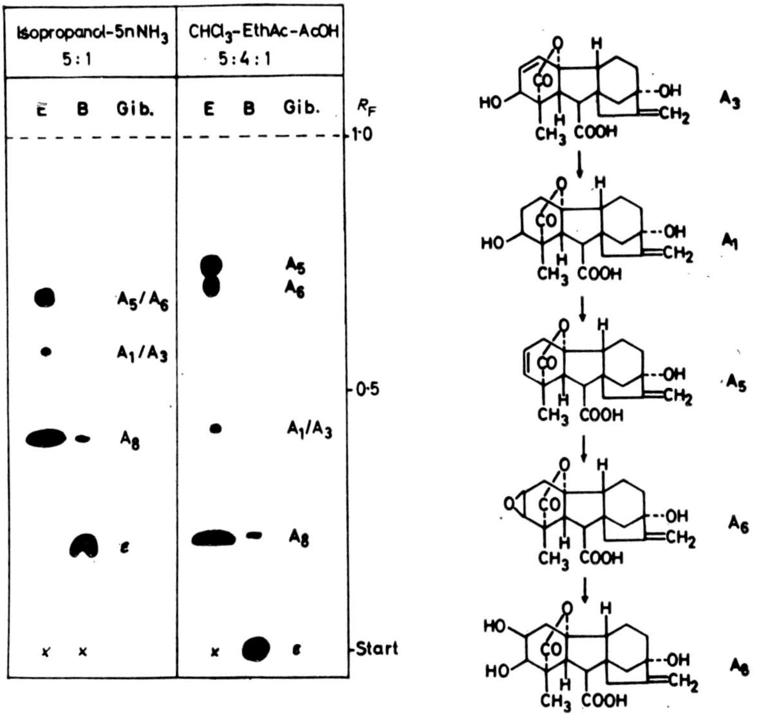

FIG. 1. *Gibberellins of bean fruits,* Phaseolus coccineus *L.*
T.l.c. with silica gel G
E: ethyl acetate extract B: n-butanol extract

Isolation and chemical structure

In order to obtain sufficient amounts of this new gibberellin '*Phaseolus* ε' for investigating its structure and its biological activity, this substance was isolated on a large scale (Fig. 2).[2] Immediately after harvesting, 3760 kg immature beans (seeds and pods) were extracted with 5000 l of 80% methanol. Each step of the separation and purification procedure was controlled by thin-layer chromatography and biological testing. Finally, in addition to small amounts of gibberellins A_1 and A_3 as well as larger amounts of A_5, A_6, and A_8, '*Phaseolus* gibberellin ε' was obtained in an overall yield of about 4×10^{-5}%. The chromatographically pure gibberellin '*Phaseolus* ε' could not be crystallised (amorphous, $[\alpha]_D + 6{\cdot}7°$ in methanol) but has been characterised by preparing a crystalline hexaacetate, m.p. 247° and $[\alpha]_D + 14{\cdot}6°$ n chloroform.

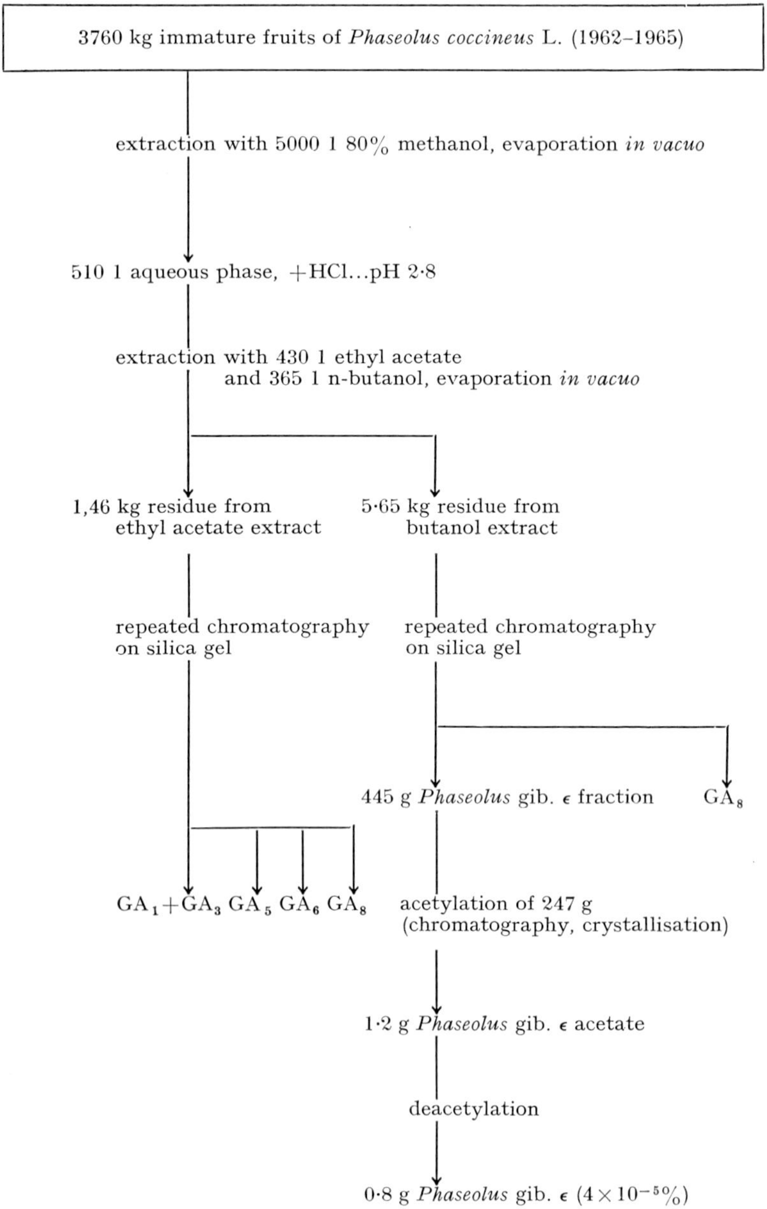

FIG. 2. *Scheme of isolation of* Phaseolus *gibberellin* ε *from immature bean fruits*

In the course of its structural elucidation, the gibberellin was hydrolysed by three different methods (Fig. 3). Thus hydrolysis with β-glucosidase at pH 5·0 in a 0·5M acetate buffer yielded 1 mole glucose and 1 mole gibberellin A_8, identified and determined quantitatively by thin-layer chromatography. Hydrolysis for 6 h at 100° with N sulphuric acid yielded 1 mole of glucose and 2 terpenoidal compounds in a ratio of 5 to 1 which can also be obtained from free gibberellin A_8 under similar conditions. The principal degradation product proved to be identical in every respect with a compound already described;[3] the structure of the other degradation product was established by means of ultra-violet, infra-red, n.m.r., and mass spectroscopy and by thin-layer electrophoresis. Hydrolysis of '*Phaseolus* gibberellin ε' with alkali yielded no glucose.

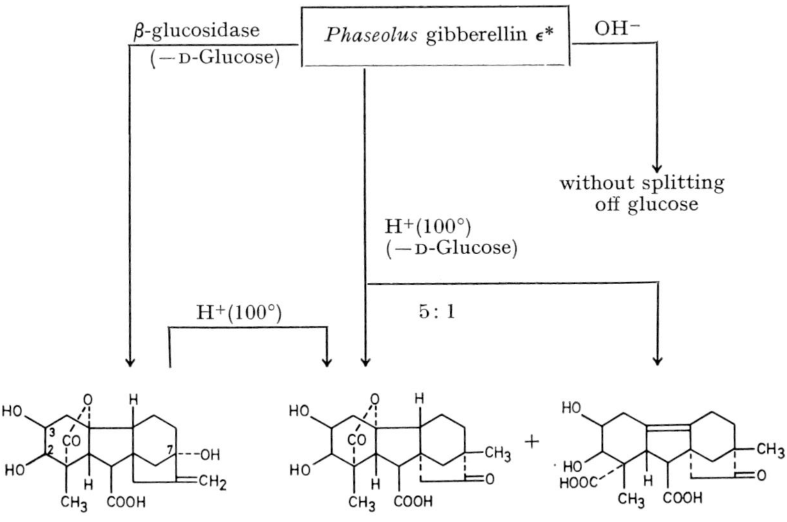

A_8 $C_{19}H_{24}O_7$†

FIG. 3. *Hydrolyses of* Phaseolus *gibberellin* ε

* amorph., $[\alpha]_D$ +6·7° in methanol; hexa-acetate (needles), m.p. 247°C, $[\alpha]_D$ +14·6° in chloroform

† m.p. 258°C, $[\alpha]_D$ −62·8° in methanol.

According to these results '*Phaseolus* gibberellin ε' is a gibberellin A_8 glucoside. In order to establish the position of attachment of the glucose moiety, the gibberellin glucoside was degraded as shown in Fig. 4.

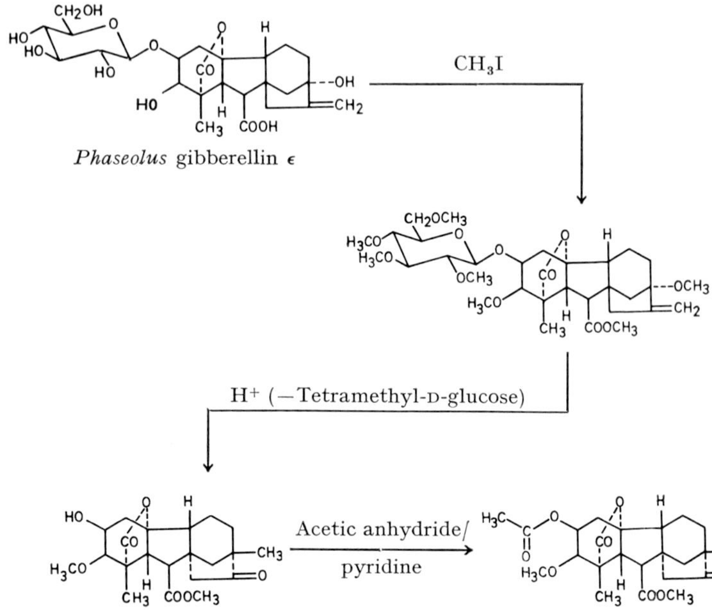

$C_{23}H_{30}O_8$*

FIG. 4. *Degradation of* Phaseolus *gibberellin* ε via *the permethylated glucoside*
*m.p. 192–193°, $[\alpha]_D$ +35·9° in methanol

These results show that '*Phaseolus* gibberellin ε' is the O(3)-D-glucopyranoside of gibberellin A_8. The glucoside was converted to the hexamethoxy methyl ester by permethylation with methyl iodide in dimethylformamide in the presence of silver oxide. Hydrolysis with 5% methanolic hydrochloric acid yielded 2,3,4,6-tetramethyl-D-glucose and a monohydroxy ketone which, by acetylation with acetic anhydride–pyridine, gave a mono-acetate the structure of which was established by use of ultra-violet, infra-red, n.m.r., and mass spectrometry. The presence of an acetoxy group at the 3-position—where the glucose was formerly attached, was unequivocally proved by the n.m.r. spectrum (Fig. 5) which shows a well resolved septet at $\delta = 4{\cdot}995$ ppm with the calculated intensities representing the axial proton at C_3

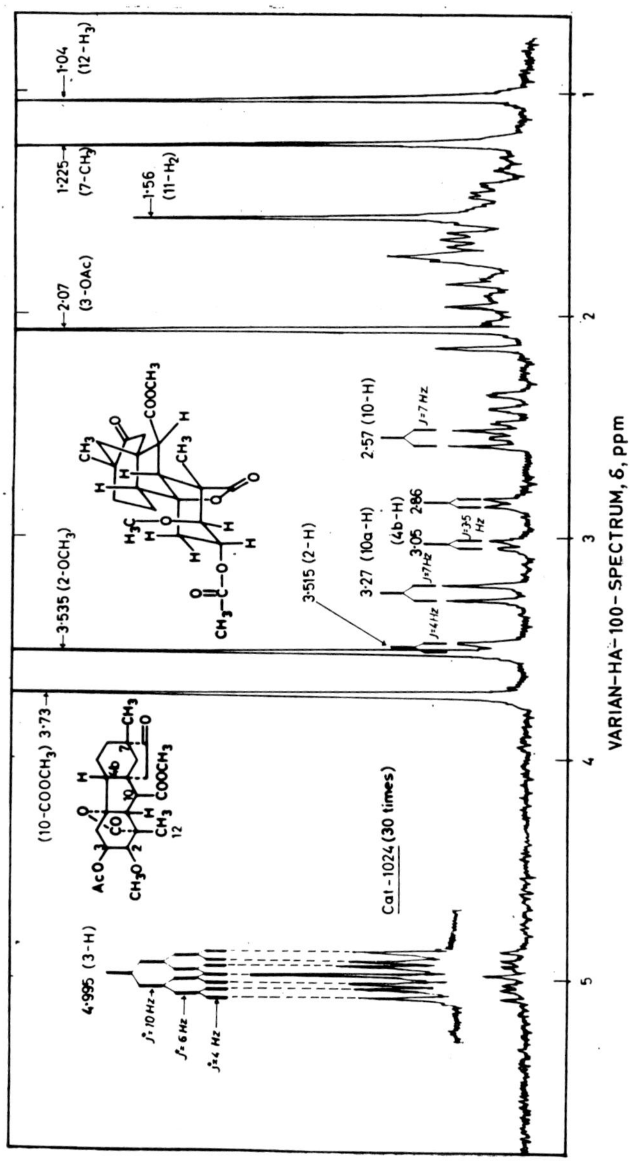

FIG. 5. *N.m.r. spectrum of a degradation product of permethylated* GA_8 *glucoside*

In the case of a 2-acetoxy group, the n.m.r. spectrum would have shown a small doublet instead of a broad septet at about $\delta = 5$ ppm.

The β-configuration is required for the glucosidic linkage because of its cleavage by β-glucosidase and this has been confirmed by the n.m.r. spectra of the gibberellin glucoside hexaacetate (doublet at $\delta = 4{\cdot}47$ ppm $J = 9$ Hz) and of the permethylated glucoside (doublet at $\delta = 4{\cdot}25$ ppm, $J = 7{\cdot}5$ Hz), as well as by the molecular rotation difference between the glucoside and its aglycone, according to the rule of Klyne.[4]

Taken together, these results show that the new polar gibberellin is the O(3)-β-D-glucopyranoside of gibberellin A_8.[5] Together with O(2)-acetyl gibberellin A_3, some time ago isolated from cultures of the fungus *Fusarium moniliforme* Sheld.,[6] it belongs to a novel group of endogenous gibberellins, which may be designated as 'conjugated gibberellins' (Fig. 6).

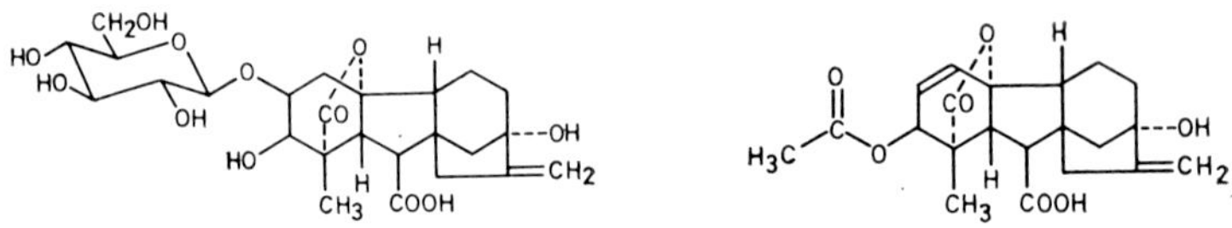

FIG. 6.

Phaseolus gibberellin ϵ= O(3)-β-D-glucopyranosyl-gibberellin A_8

O(2)-Acetyl-gibberellin A_3

Metabolism and function

During the ripening process in beans, the content of gibberellin A_8 glucoside increases and the content of non-conjugated gibberellins decreases. Finally, in mature dry seeds, only the glucoside is present. Table I summarises the data on the distribution of gibberellins in green pods, immature and ripe seeds and in cotyledons from young seedlings, as determined by thin-layer chromatography and dwarf pea bioassay. These results suggest, that the GA_8 glucoside may be an end product of the metabolism of gibberellins during seed ripening. Studies with isotopically labelled gibberellins A_3 and A_6 confirm, that GA_8 glucoside readily arises from non-conjugated gibberellins.[2,7] Thus, tritiated gibberellin A_6 applied to ripening bean fruits was partly transformed into labelled gibberellin A_8 glucoside, although radioactivity was also found in the GA_1/GA_3 and to a higher extent in the GA_8 fraction (Fig. 7). The [^{3}H]GA_6 was prepared by Wilzbach tritiation of GA_6 followed by repeated chromatography on silica gel and crystallisa-

Table I

Occurrence of gibberellins in different parts of bean fruits

Gibberellins	green pods	immature seeds	ripe seeds	germinating seeds	cotyledons of seedlings
A_1/A_3	+	+		+	+
A_5/A_6	+	+		+	+
A_8	+	+		+	+
A_8 glucoside	+		+	+	+

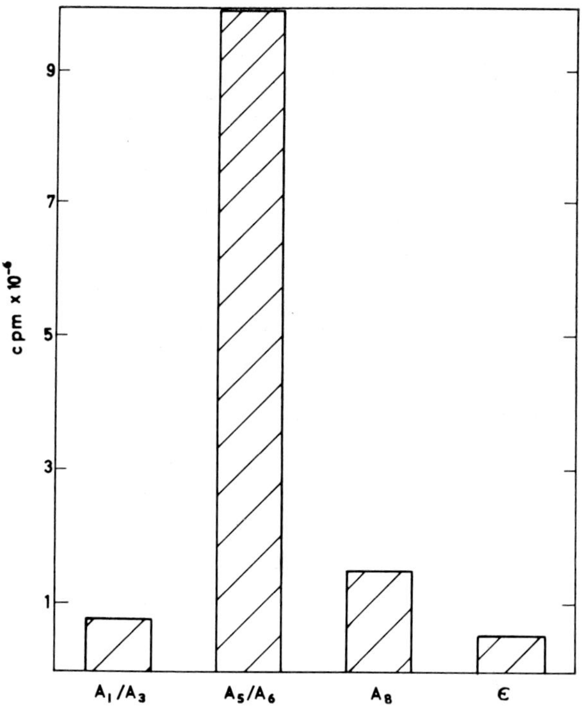

FIG. 7. *Application of [^3H]gibberellin A_6 to ripening bean fruits*
Pods after 5 days' incubation

tion. Furthermore, uniformly labelled [^{14}C]gibberellin A_3 synthesised from [^{14}C]acetate by *Fusarium moniliforme* Sheld. was applied to maturing bean fruits, and after a one and two days incubation, respectively, the pods and seeds were extracted in the usual manner. Most of the radioactivity was found in the pods, a smaller part also in the seeds. The pods contained, in addition to unchanged [^{14}C]gibberellin A_3, three more polar radioactive substances, which could be extracted with n-butanol from the acidified aqueous solution, previously extracted exhaustively with ethyl acetate (Fig. 8). Fraction I is not yet identified. Fraction II

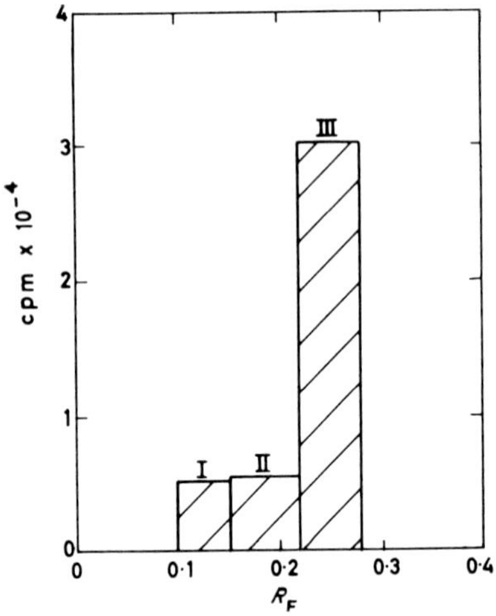

FIG. 8. *Application of [^{14}C]gibberellin A_3 to ripening bean fruits*
n-Butanol phase after 2 days' incubation
T.l.c.: silica gel G; n-propanol/5N NH_3

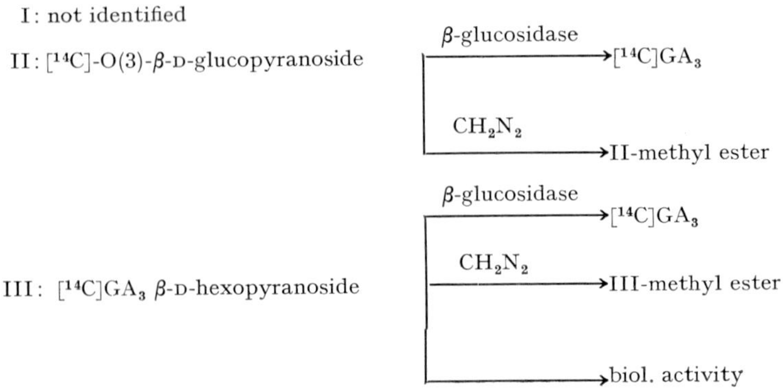

proved to be gibberellin A_8 glucoside, as shown by hydrolysis with β-glucosidase yielding [^{14}C]GA_8 and comparison of this compound, the radioactive glucoside and its methyl ester with authentic compounds using thin-layer chromatography. Fraction III, which arose in larger amounts after only 1 day incubation was identified as a gibberellin-A_3-hexopyranoside in the following manner: (1) cleavage by β-glucosidase yielding [^{14}C]gibberellin A_3, (2) methylation with diazomethane leading to a radioactive methyl ester, which proved to be chromatographically identical with non-labelled O(2)-β-D-glucopyranosyl-gibberellin A_3 methyl ester, synthesised from gibberellin A_3 and acetobromoglucose[8] in our laboratory[2,9] and (3) detection of a high biological activity of the same order as that of gibberellin A_3 for the radioactive glucoside.

In this connexion it should be mentioned that several years ago, Murakami[10] described an *in vitro* formation of a gibberellin A_3 glucoside, identified by paper chromatography, by incubation of GA_3 and glucose with cucumber leaf discs. Very recently, O(2)-glucopyranosyl–gibberellin A_3 has been isolated from *Pharbitis* seeds.[11,12]

Function in plant metabolism

The natural occurrence of GA_8 glucoside in beans and of GA_3 glucoside in *Pharbitis* as well as the rapid formation of GA_3 glucoside in bean pods after application of GA_3 suggest that these glucosidic forms may possess distinct functions in the gibberellin metabolism of plants. In order to clarify these possible functions, the biological activity of the GA_8 glucoside has been tested in different bioassays.[7] In the dwarf pea bioassay under red light conditions, the gibberellin A_8 glucoside is of low but significant activity, of the same order as gibberellin A_8 (Fig. 9). In the lettuce hypocotyl test both GA_8 and its glucoside are inactive (Fig. 10). With maize dwarf mutants, there were differences in biological activity depending on whether the plants were illuminated or not. Thus, the mutant d-5, grown in the light, shows only low sensitivity to GA_8 and shows no response to GA_8 glucoside (Fig. 11). In the dark, however, the same mutant is much more sensitive to GA_8, but remains inactive to the glucoside (Fig. 12). From these results it may be concluded that gibberellin A_8 glucoside and perhaps also gibberellin A_8 are deactivated forms of gibberellin, produced from biologically active gibberellins during seed ripening.

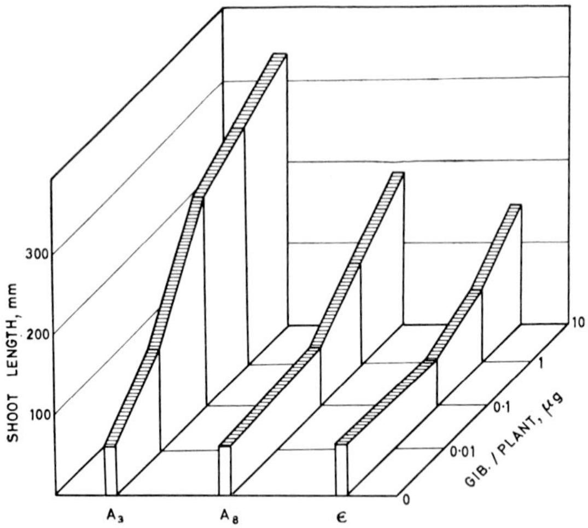

FIG. 9. *Comparative potency of gibberellins A_3, A_8, and A_8 glucoside in the dwarf pea bioassay* *Pisum sativum* L. c.v. 'Meteor'

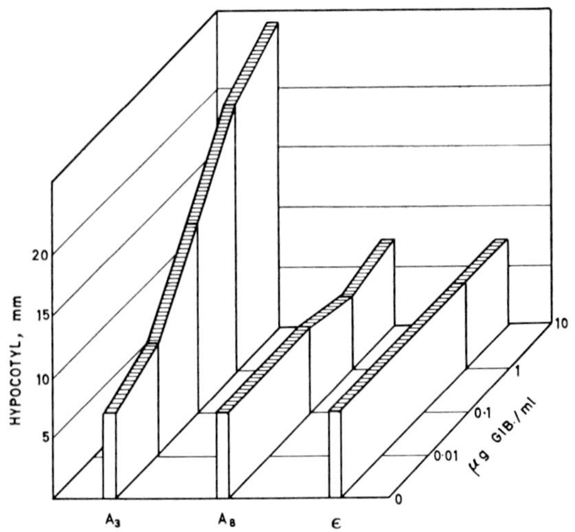

FIG. 10. *Comparative potency of gibberellins A_3, A_8, and A_8 glucoside in the lettuce hypocotyl test* *Lactuca sativa* L. c.v. 'Grand Rapids'

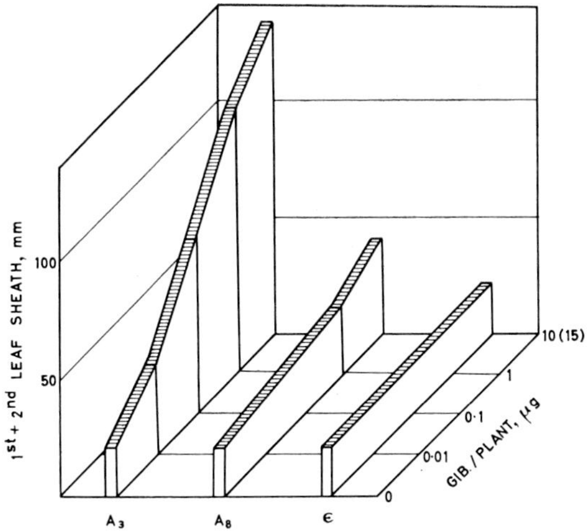

FIG. 11. *Comparative potency of gibberellins A_3, A_8, and A_8 glucoside in the dwarf maize bioassay in the light* *Zea mays* L. (d-5)

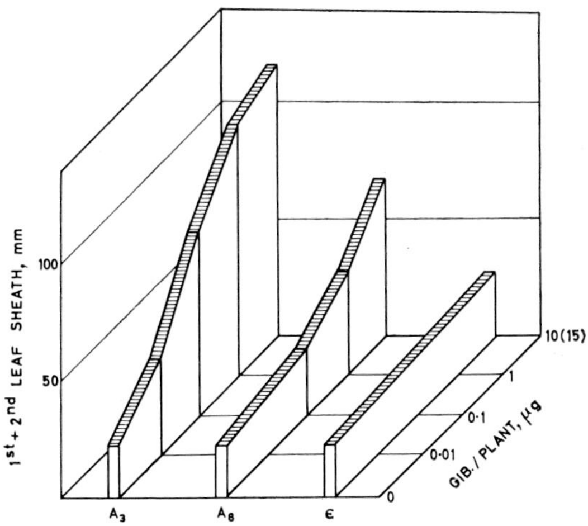

FIG. 12. *Comparative potency of gibberellins A_3, A_8, and A_8 glucoside in the dwarf maize bioassay in the dark* *Zea mays* L. (d-5)

The next question is, whether gibberellin A_8 glucoside is reconverted into more active gibberellins during seed germination. If it is so converted, the glucoside could be a depot gibberellin. To investigate this possibility, two types of experiments were carried out.[7] Firstly, bean seeds were incubated with AMO-1618,* an inhibitor of the gibberellin biosynthesis at the stage of cyclisation of geranyl–geranyl pyrophosphate. After germination the gibberellin content of the cotyledons was estimated using thin-layer chromatography and quantitative dwarf pea bioassay (Table II). Normally, a cotyledon of dark-germinated bean seedlings contains about 0·20 ng GA_3 equivalents of fraction GA_1/GA_3, 0·25 ng GA_5, 0·50 ng GA_6, 0·12 ng GA_8, and 0·18 ng GA_8-glucoside. After AMO application, cotyledons of germinated seeds lack GA_1/GA_3, GA_5, and GA_6, but the GA_8 and GA_8 glucoside contents are the same as those in water–treated seeds. From these results it may be concluded that, during germination, gibberellin is synthesised from precursors, located before geranyl–geranyl pyrophosphate cyclisation in the biogenetic scheme, but it is uncertain whether AMO-1618 inhibits only this latter reaction. It seems possible that other metabolic stages may be involved, perhaps the GA_8 conversion into more active gibberellins. It

Table II

Gibberellin content in cotyledons of bean seedlings (7d, dark)

	GA_3, equiv./cotyledon, ng*				
Seeds 48 h in:	GA_1/A_3	GA_5	GA_6	GA_8	GA_8 glucoside
H_2O	0·20	0·25	0·50	0·12	0·18
$5\cdot10^{-5}$M AMO-1618	0	0	0	0·15	0·12

*Quantitative dwarf pea bioassay after t.l.c.

should be mentioned, that Paleg reported the inhibition of steroid biosynthesis by AMO-1618.[13] Therefore, the present results with this inhibitor allow no decision whether GA_8 glucoside has a depot function or not, because it cannot be excluded that a GA_8 conversion occurs during germination, either in addition to gibberellin biosynthesis or without it.

* 2-isoprophyl-4-dimethylamino-5-methylphenyl 1-piperiline carboxylate methyl chloride.

The second series of experiments involved the use of labelled gibberellin A_8 glucoside. After application of [^{14}C]gibberellin A_3 to maturing bean fruits, ripe seeds were harvested and investigated with regard to the occurring radioactive gibberellins, one part of the seeds before and the other part after germination. The ethyl acetate and butanol fractions obtained showed different amounts of radioactivity in ungerminated and in germinated seeds:

	c.p.m.	
	ethyl acetate fraction	n-butanol fraction
Ungerminated	$0{\cdot}9 \times 10^4$	$1{\cdot}9 \times 10^4$
Germinated	$1{\cdot}8 \times 10^4$	$0{\cdot}6 \times 10^4$

This result may be easily explained as the reconversion of the polar, butanol soluble gibberellin glucoside into ethyl acetate soluble, non-conjugated gibberellins during seed germination. But the definite decision whether GA_8 glucoside is a depot gibberellin or not, can only be made after further investigations.

Another possible function of GA_8 glucoside in gibberellin metabolism is concerned with the translocation of these growth regulators. Investigations on the gibberellins in bleeding sap of trees[14] has led to the conclusion that gibberellin glycosides may be a favoured form for translocation. Bleeding sap of Norway maple, *Acer platanoides* L., (170 ml) and Wych elm, *Ulmus glabra* Huds., (300 ml) collected in the early spring have been found to contain mainly polar gibberellin-like substances which could be extracted from the acidified sap with n-butanol. β-Glucosidase split off less polar, ethyl acetate soluble, compounds with gibberellin activity, which could be identified by use of thin-layer chromatography as gibberellin A_8 in sap of maple and as gibberellin A_3 (A_1) in sap of elm, respectively. For more detailed studies, 10 l of bleeding sap of Wych elm were collected and worked up in the manner shown in Fig. 13. After acidifying to pH 2·8 the sap was extracted, with ethyl acetate(×5), and then with n-butanol (×3). Thin-layer chromatography and bioassay of the ethyl acetate fraction showed one active zone at R_f 0·05–0·26. Hydrolysis with β-glucosidase, followed by ethyl acetate extraction, thin-layer chromatography, and bioassay yielded a gibberellin active compound with the same R_f as gibberellin A_3. The residue of the

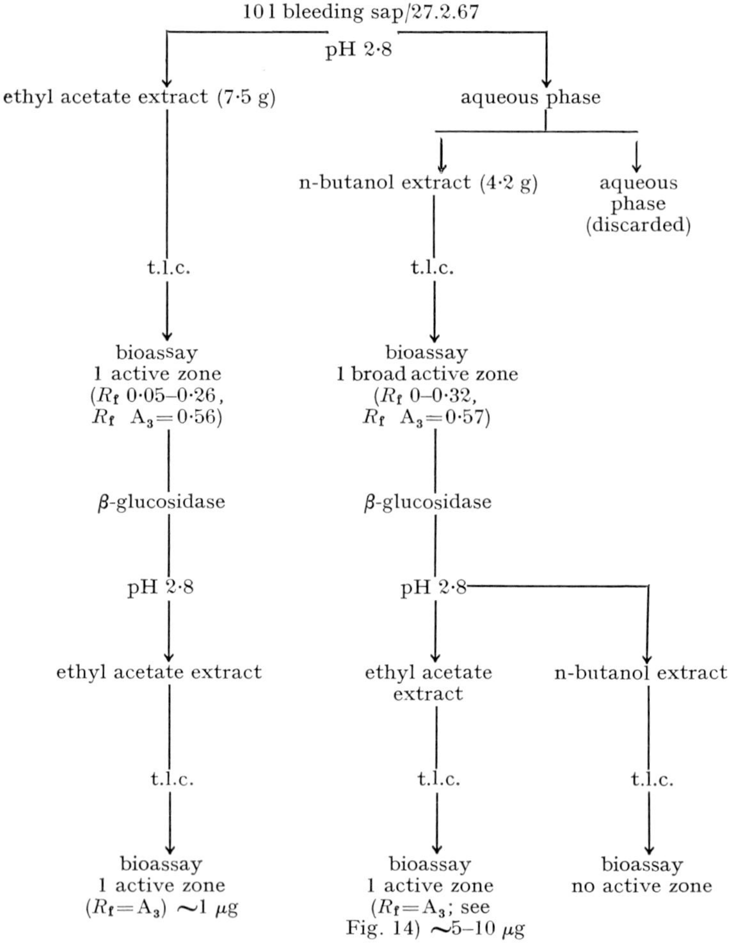

FIG. 13. *Investigations of the gibberellins in elm bleeding sap Ulmus glabra* Huds.

butanol fraction has been also chromatographed on thin-layers of silica gel and tested for biological activity. Gibberellin-like activity was found in zones between the origin and R_f 0·32. Hydrolysis with β-glucosidase and acidification was followed by extractions with ethyl acetate and n-butanol. The butanol extract no longer contained gibberellin activity, but in the ethyl acetate fraction gibberellin A_3 could be identified by repeated thin-layer chromatography using different solvent systems. One part of the chromatographic investigation is summarised in Fig. 14. After thin-layer chromatography with kieselguhr only the GA_3 zone, which was accurately separated from GA_1, showed biological activity in the dwarf pea bioassay. These results demonstrate that the sample of elm bleeding sap investigated apparently contained only one polar gibberellin-like substance which proved to be a gibberellin A_3 hexopyranoside.

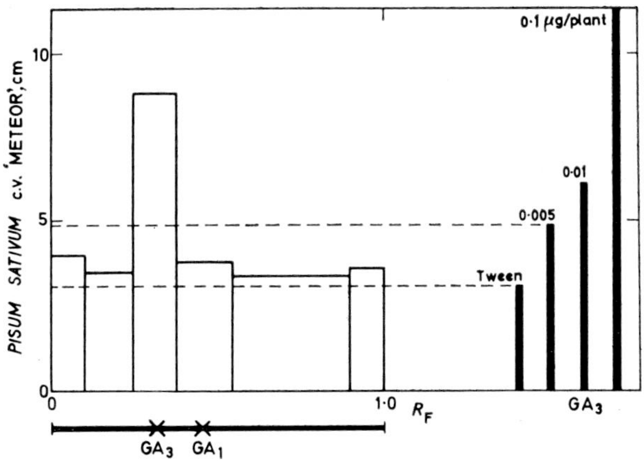

FIG. 14. *Combined thin-layer chromatography and gibberellin bioassay of one fraction of elm bleeding sap* (*see Fig.* 13) *Ulmus glabra* Huds.
T.l.c.: Kieselguhr G; benzene/H_2O/propionic acid

Conclusion

These studies are the first steps in a detailed chemical analysis of gibberellin translocation in bleeding sap. The occurrence of gibberellin-like substances in bleeding sap of *Helianthus, Lupinus, Pisum, Impatiens,* and *Vitis* has been reported by other workers[15–21] but the identification and chemical characterisation of these substances have still to be completed.

References

1 Sembdner, G., Schneider, G., Weiland, J., & Schreiber, K., *Experientia,* 1964, **20,** 89

2 Weiland, J., Diss. Univ. Halle, 1968

3 MacMillan, J., Seaton, J. C., & Suter, P. J., *Tetrahedron,* 1962, **18,** 349

4 Klyne, W., *Biochem. J.,* 1950, **47,** XLI

[5] Schreiber, K., Weiland, J., & Sembdner, G., *Tetrahedron Lett.*, 1967, p. 4285
[6] Schreiber, K., Schneider, G., Sembdner, G., & Focke, I., *Phytochemistry*, 1966, **5,** 1221
[7] Sembdner, G., Weiland, J., Aurich, O., & Schreiber, K., in preparation
[8] Koenigs, W., & Knorr, E., *Ber. dt. chem. Ges.*, 1901, **34,** 957
[9] Schreiber, K., Weiland, J., & Sembdner, G., in preparation
[10] Murakami, Y., *Bot. Mag., Tokyo,* 1961, **74,** 424
[11] Tamura, S., Takahashi, N., Murofushi, N., Yokota, T., & Kato, J., 6th Int. Conf. Plant Growth Substances, Ottawa, 1967
[12] Tamura, S., Takahashi, N., Yokota, T., & Murofushi, N., *Planta,* 1968, **78,** 208
[13] Paleg, L. G., 6th Int. Conf. Plant Growth Substances, Ottawa, 1967
[14] Sembdner, G., & Schreiber, K., in preparation
[15] Carr, D. J., Reid, D. M., & Skene, K. G. M., *Plantax* 1964, **63,** 382
[16] Phillips, I. D. J., & Jones, R. L., *Planta,* 1964, **63,** 269
[17] Jones, R. L., & Phillips, I. D. J., *Pl. Physiol., Lancaster,* 1966, **41,** 1381
[18] Jones, R. L., & Phillips, I. D. J., in preparation
[19] Reid, D. M., & Carr, D. J., *Planta,* 1967, **73,** 1
[20] Skene, K. G. M., *Planta,* 1967, **74,** 250
[21] Sitton, D., Richmond, A., & Vaadia, Y., *Phytochemistry,* 1967, **6,** 1101

DISCUSSION

Professor J. W. Fairbairn.—Since the glycosides are more water soluble than the free gibberellin, would this justify the conclusion that the glycosides represent a transport form of the gibberellins?

Dr. Sembdner.—Yes. The increased water solubility of gibberellin glycosides does agree with their possible function as a translocation form in the bleeding sap of trees. But in addition to this, the glycosylation represents a deactivation, which may also be necessary for translocation.

Professor B. B. Stowe.—Would you comment on similarities and differences between your *Phaseolus* species and that used by Dr. J. MacMillan?

Dr. Sembdner.—To my knowledge *Phaseolus multiflorus* Lam. is a synonym for the correct name *Phaseolus coccineus* L. But I do not know if Dr. MacMillan uses the same variety as we do.

Dr. J. van Bragt.—What is the situation on gibberellins in the bleeding sap of tomato roots? Is it as simple as you described for *Acer* and *Ulmus*?

Dr. Sembdner.—Our studies have not been finished yet, but with tomatoes the situation seems to be much more complicated, because apparently more forms of gibberellins are involved. Especially in tomato bleeding sap the amount of gibberellins is very low and of great variance, possibly according to environmental conditions. Therefore we are now growing the plants in nutrient solution in a climatic chamber.

HORMONES IN THE XYLEM SAP OF APPLE TREES

By L. C. LUCKWILL and PAMELA WHYTE

(*Long Ashton Research Station, University of Bristol*)

Xylem sap extracted by suction from branches of apple in spring was found to contain two major hormones which could be separated by ascending paper chromatography in water. One was a cytokinin (R_f 0·67–0·70) active in delaying senescence of isolated segments of oat leaves and in stimulating cell division in soybean cotyledon callus. The other was a highly polar compound (R_f 0·85–0·90) showing activity in the *Avena* mesocotyl test. This compound was inactive in the lettuce hypocotyl, barley endosperm and *Rumex* leaf senescence tests for gibberellin, the soybean callus test for cytokinin, and in the *Coleus* petiole abscission test for auxin.

Both hormones appeared in the sap in early spring and reached their maximum concentration about the time of full bloom, after which they decreased in concentration. No hormone activity could be detected in sap collected during the period from late September to January.

In addition to the two major hormones, traces of gibberellin activity were sometimes found in sap samples collected in early spring.

Complete removal of buds and extension growth from branches on 1st April did not decrease the amount of either hormone found in the xylem sap on 9th May. From this evidence and from the fact that similar activity was found in bleeding sap collected from birch, sycamore and field maple trees near ground level, it is concluded that these hormones have their origin in the root system. The possible significance of xylem sap hormones in relation to shoot growth is discussed.

Introduction

THE xylem has long been known as the chief route by which water, mineral salts, and—in woody plants—reserve carbohydrates and soluble nitrogenous compounds are translocated from the root to the shoot. It is only recently, however, that its importance as a pathway of hormone translocation has been recognised.

As early as 1943, Went,[1] on the basis of experiments with tomato, postulated the existence of a hypothetical hormone 'caulocaline' which he believed to be produced in the roots and to be responsible for shoot growth and leaf expansion. In 1954, Chibnall & Wiltshire[2] showed that the presence of roots on runner bean seedlings was essential for the maintenance of protein synthesis in the leaves, but at that time nothing was known concerning the nature of the hormones involved. Interest in this topic was revived following the discovery by Richmond & Lang[3] in 1957 of the effect of kinetin (6-furfurylaminopurine) on protein synthesis in detached leaves, and the subsequent finding by Kulaeva[4] that the xylem sap of *Nicotiana rustica* contained a substance that would retard the senescence of detached leaves in the same manner as kinetin. In 1964, when the present investigation was started, Loeffler & van Overbeek[5] reported a similar cytokinin-like substance in the bleeding sap of the grape vine, and Kende[6] found two substances on paper chromatograms of root exudates of *Helianthus annuus* that would retard chlorophyll degradation in barley leaf sections. One of the substances from sunflower sap was active in the soybean callus assay[7] regarded as specific for cytokinins. Since 1964, further evidence of cytokinins in bleeding sap or in extracts of roots of a number of herbaceous and woody species has been reported by Itai & Vaadia,[8] Nitsch & Nitsch,[9] Seth & Wareing,[10] Weiss & Vaadia,[11] Carr & Burrows,[12] and Skene & Kerridge.[13] Little work has been done with apple, except that of Jones,[14] who reported that xylem sap extracted in May from apple rootstock '3430' was active in promoting cell division in cultures of tobacco pith, but not in promoting the growth of apple shoot tips in sterile culture.

In addition to cytokinins, other hormonal substances have been reported to occur in xylem sap. Carr *et al.*[15] detected gibberellin activity and inhibitory substances in bleeding sap of *Lupinus albus*, *Pisum sativum* and *Impatiens glandulifera*; Phillips & Jones[16] found gibberellin-like activity in the bleeding sap from root systems of *Helianthus annuus*; Skene[17] found a gibberellin-like substance in the root exudate of grape; and Nitsch & Nitsch[9] in the sap collected from sugar maple found precursors which were converted to active auxin-like substances on autoclaving.

The present work was undertaken to discover what types of hormones occurred in the xylem sap of apple and how their concentrations varied throughout the season. It forms part of a wider investigation into the hormonal physiology of the apple tree.

Experimental

Trees used

The 1964–65 series of sap extractions were made from a family of open pollinated seedlings of the late-flowering cider apple 'Stoke Red'. These ungrafted trees, part of a breeding project, were 17 years old in 1964 and all but a few trees had outgrown their juvenile phase. They were all relatively tolerant of apple scab and received no pest and disease control sprays. Weed control around the tree bases in 1964 and 1965 was by means of paraquat and simazine (1 and 1·5 lb active per acre respectively) applied in March.

The 1966 collections of sap were made from 13-year-old bush trees of apple 'Cheddar Cross' on M.II rootstocks. The trees were exceptionally vigorous but carried only a light crop in 1966. They received a full spray programme, and weed control was by means of paraquat at 0·75 lb/acre.

The experiment on the effect of dis-budding was carried out in 1966 on 10-year-old cropping trees of apple 'Stoke Red'. Pest and weed control was the same as for 'Cheddar Cross'.

Sap extraction

At intervals throughout the season, branches 3 to 4 ft long and from 0·5 to 1 in. in diameter at the base, were collected at random from the experimental trees. Xylem sap was extracted into a Buchner flask under reduced pressure using the technique developed by Bollard.[18] The crude sap was placed in polythene bags and stored in the frozen condition at $-20°$ until required for assay. A minimum of 100 ml of sap was extracted on each occasion. The sap was concentrated 15 times under vacuum in a rotary evaporator at 40° before chromatography.

Chromatography

Ascending paper chromatography on Whatman No. 1 paper was used to separate the various constituents of the sap before bioassay. A number of different solvent mixtures were used, but water was found to be the most generally useful, and was employed for the main experiment on seasonal variation in hormone content. Chromatograms were cut into 21 portions (0–20) and each section was assayed biologically. The 't' test was used to assess the statistical significance of the results, only those 'peaks' of activity which exceeded the 5% confidence limits being considered significant.

Bioassay methods

Preliminary work carried out with bleeding sap from birch, sycamore and field maple showed that, in addition to cytokinins, xylem sap contained appreciable quantities of gibberellin and possibly an auxin-like substance as well. For this reason a rather wide range of bioassay methods were tried in the early stages of the work. These included the lettuce hypocotyl test,[19] the barley endosperm test[20] and the *Rumex* leaf senescence test,[21] all of which are specific for gibberellins; the soybean callus assay[7] and a test based on the senescence of sections of *Avena* leaves for cytokinins; the *Coleus* petiole abscission test[22] for auxins; and the *Avena* mesocotyl test[23] which reacts to auxins and gibberellins, but not to cytokinins.

Eventually, the soybean callus test and the *Avena* mesocotyl test were selected as the most suitable for routine work with apple sap hormones. For the soybean callus test, amounts of sap concentrate corresponding to 1·5 and 7·5 ml crude sap were spotted on the chromatograms. After development in water, the appropriate R_f zones were cut out and added to 10 ml of the basic medium, less kinetin. The test was carried out essentially as described by Miller,[7] the fresh weight of callus being measured after a period of four weeks in darkness at 25°. Control cultures containing known amounts of kinetin were included with each assay. Since it was necessary to add the chromatogram segments to the culture medium before sterilisation, a separate experiment was carried out to investigate the possible effect of autoclaving on the cytokinin activity of natural sap. Samples of crude apple sap, together with partially-purified cytokinins extracted from birch, sycamore and field maple sap with n-butanol, were divided into two equal portions. One portion was sterilised by autoclaving at 15 lb pressure for 20 minutes; the other was sterilised by membrane filtration. Subsequent assay in the soybean callus test confirmed that there was no decrease, or increase, in activity as the result of autoclaving.

The *Avena* mesocotyl test was carried out as described by Nitsch & Nitsch[23] using huskless 'Brighton' oats. For this test, the sap concentrate was spotted on the starting line of the chromatograms in amounts corresponding to 3·0, 0·3 and 0·03 ml crude sap. After development in water, the appropriate R_f zone was cut out and placed directly in the tube containing the mesocotyl sections and sucrose/buffer. Since the duration of the test was only 22 h, sterilisation of the medium was not required. The activity of the sap was measured against IAA standards. Both the soybean callus assay and the mesocotyl test were run in duplicate, using two replicate chromatogram strips.

Results

The two main types of hormone activity which were found in the xylem sap of apple are designated Substance K and Substance A.

Substance K was active in stimulating cell division in soybean callus and in retarding the senescence of isolated sections of oat leaves (cv. 'Garry'). It showed no activity in any of the other bioassay systems used. For this reason, it is believed to be a cytokinin. When chromatographed in water, substance K formed a compact spot with a peak concentration around R_f 0·67–0·70 (Fig. 1). The highest concentration of this substance found in apple sap was of the order of 200 μg equivalents of kinetin per litre (0·2 ppm). In apple seedling sap collected in April, some cytokinin activity was present also at R_f 0·52 on the water chromatograms. Although

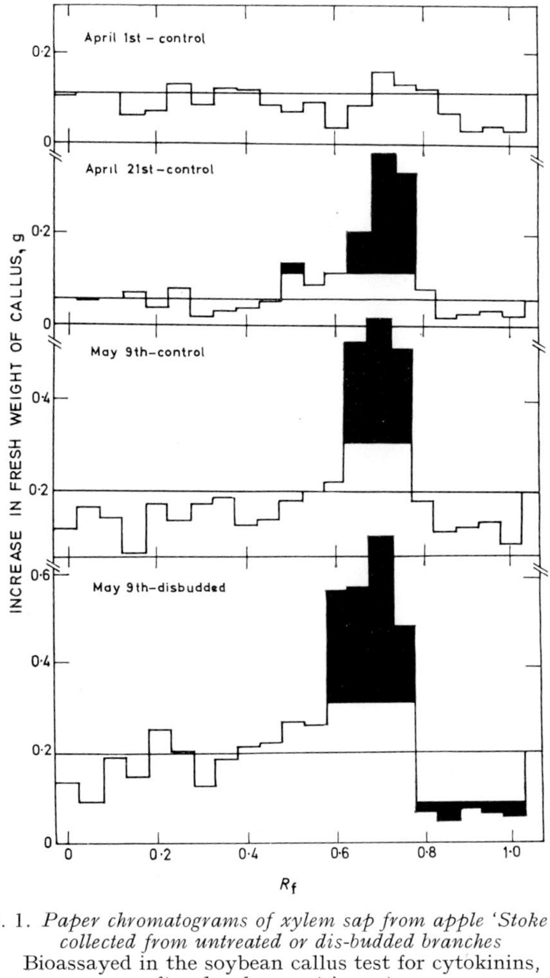

FIG. 1. *Paper chromatograms of xylem sap from apple 'Stoke Red' collected from untreated or dis-budded branches*
Bioassayed in the soybean callus test for cytokinins, after development in water.

this second spot could have been an artefact due to trailing of the main peak, the presence of a second cytokinin of lower R_f seems to be indicated. The maximum concentration of this R_f 0·52 compound was equivalent to 76 μg kinetin per litre. This second compound was found in the apple seedling sap only in April. It was also present in small quantity in sap from the 'Stoke Red' trees collected on 21st April, 1966, (Fig. 1), but was not detected in 'Cheddar Cross'. In Figs. 4 and 5, the concentrations of both compounds have been added to give a total cytokinin concentration.

Samples of 'Cheddar Cross' sap containing high cytokinin activity were combined and adjusted to pH 7. On partitioning this sap three times with an equal volume of methylene dichloride or ethyl acetate, all the activity remained in the aqueous layer. When shaken with n-butanol, however, the cytokinin activity was completely removed from the aqueous phase and was recovered from the butanol.

Substance A was detected by its activity in the *Avena* mesocotyl test. At first, it was thought that it might be a gibberellin, but this possibility seemed to be ruled out when it was found to be completely inactive in the lettuce hypocotyl test, the *Rumex* leaf senescence test and the barley endosperm test. In the *Coleus* test it induced a slight acceleration, rather than a delay, of petiole abscission, which makes it doubtful whether it can be classified as an auxin either. In the mesocotyl test, it caused increases in growth of up to 350% of the controls, which makes it unlikely that it is merely an auxin synergist. Its chemical nature at present remains obscure.

Substance A is highly water soluble, slightly soluble in absolute methanol, but insoluble in acetone, n-butanol, ethyl acetate and methylene dichloride. On chromatograms developed in water it ran almost to the solvent front (Fig. 2), but in an isopropanol/water (10/1) mixture it hardly moved from the starting line. It was readily eluted with water from a silica gel column. Substance A was subjected to paper electrophoresis for 6 h in a 0·01 M dipotassium monohydrogen phosphate buffer of pH 6·5. at 300V and 2 mA. Subsequent bioassay of the electrophoresis strip showed that the compound had split into cationic and anionic portions.

Gibberellins

A systematic search for gibberellins was carried out by testing paper chromatograms of apple sap in the *Rumex*, lettuce hypocotyl

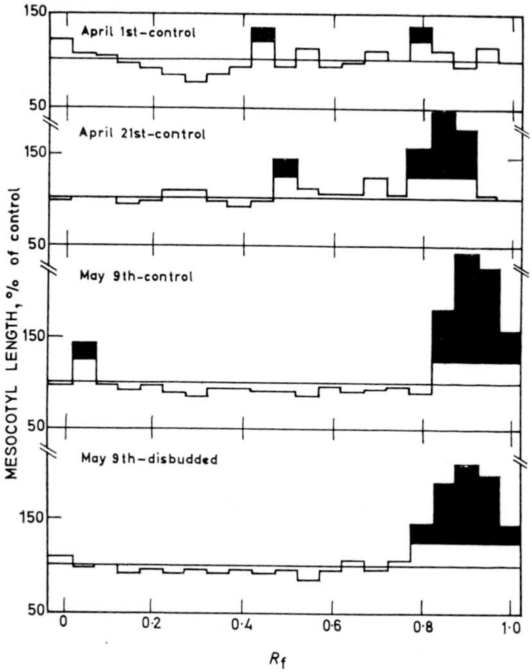

FIG. 2. *Paper chromatograms of xylem sap from apple 'Stoke Red' collected from untreated or dis-budded branches*
Bioassayed in the *Avena* mesocotyl test after development in water.

and barley endosperm tests. Using the *Rumex* test, approximately 0·01 μg equivalents GA_3/1 (0·00001 ppm) was detected in apple seedling sap collected in April 1964, and approximately 0·02 μg equivalents GA_3/l (0·00002 ppm) in sap collected from 'Stoke Red' trees on 1st April, 1966. The latter was also active in the barley endosperm test. No gibberellin activity could be detected at other times of the year and none was found in apple seedling sap collected in 1965 or in 'Cheddar Cross' sap collected in 1966. The apple sap gibberellin was a rather polar compound, having an R_f value of 0·1–0·3 in isopropanol/water (10/1) and 0·8–0·9 in water (Fig. 3).

Seasonal variation in hormone content of xylem sap

The seasonal variations in the concentrations of Substances K and A are shown in Figs. 4 and 5. In the apple seedlings both the cytokinin and Substance A first appeared in the sap either at or just before the time of bud swelling in March and reached their

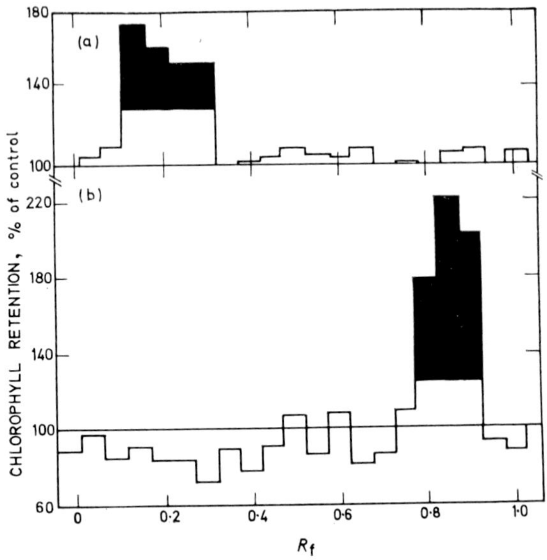

FIG. 3. *Gibberellin bioassay of paper chromatograms of xylem sap from apple 'Stoke Red', collected* 1*st April,* 1966
Rumex leaf disc senescence test.
(a) chromatogram developed in isopropanol/water (10:1).
(b) chromatogram developed in water.

optimum concentrations at blossom time in mid-May. After bloom, the concentrations of both hormones fell, reaching a very low value by early August and disappearing entirely by October. No activity could be detected in sap extracted during the winter months (Fig. 4).

The same general picture is presented by the diagrams for 'Cheddar Cross' (Fig. 5) based on extractions made in 1966, although the optimum concentrations of both hormones in this variety were less than in 'Stoke Red' seedlings sampled in 1964 and 1965. In 'Cheddar Cross' the highest concentrations of Substances K and A occurred over the period from green-cluster (30th March) to early fruitlet stage (beginning of June), with a marked drop in concentration around the time of full bloom (end of April–early May). The reason for this temporary fall in concentration is not definitely known, but it is suggested that it may be associated with the exceptionally heavy rainfall in April, 1966. This was 4·12 in. (164% of average) compared with 1·31 in. for April, 1965 and 1·77 in. for April, 1964. It is possible that rapid intake of water by

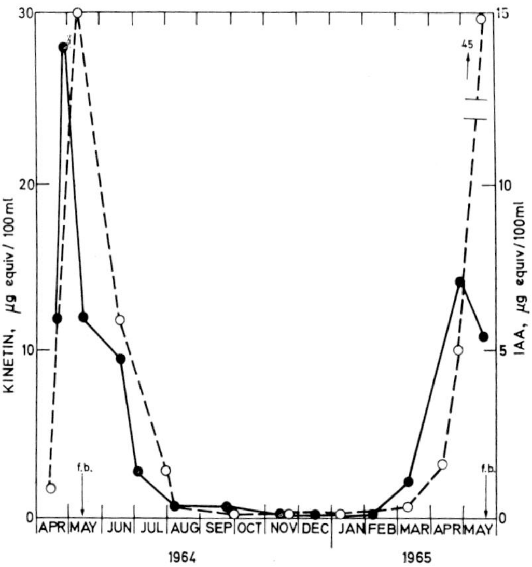

FIG. 4. *Seasonal variation in concentration of hormones K and A in the xylem sap of seedling apple trees*
f.b. = full bloom.
●——● K as μg. equivalents kinetin
○ - - - ○ A as μg. equivalents IAA

the trees during this period may have led to some dilution of the sap constituents.

The other feature of interest is the somewhat slower falling off in concentration of Substance K and A in 'Cheddar Cross' as compared with the seedling trees.

Effects of dis-budding

Since developing buds are known to produce growth-promoting substances it was thought possible that the increasing amounts of Substances K and A in the xylem sap between bud-break and full bloom might have their origin in the lateral and terminal buds, rather than in the root system. Although such substances would presumably be translocated out of the buds in the phloem, the possibility of lateral transport of growth regulators between phloem and xylem is well established. To investigate this possibility an experiment was carried out in the spring of 1966, using mature fruiting trees of the cider apple 'Stoke Red' on Crab C rootstock.

On 1st April, eight major branches on each of two trees were selected. Four branches from each tree were removed for sap

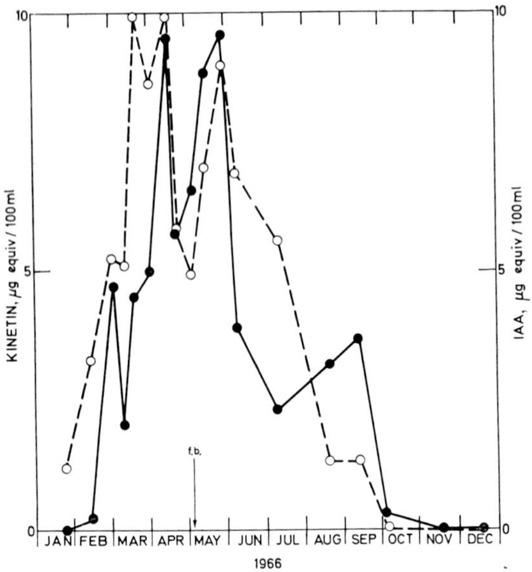

FIG. 5. *Seasonal variation in concentration of hormones K and A in the xylen sap of apple 'Cheddar Cross'*
f.b. = full bloom.
●——● K as μg. equivalents kinetin
○ - - - ○ A as μg. equivalents IAA

extraction. The other four were completely dis-budded and, in addition, had all one year old extension growth removed. Chromatograms of xylem sap showed that, at this time, no appreciable quantities of either Substance K or Substance A were present. A further check on adjacent untreated control trees on 21st April showed that both mesocotyl and cytokinin activity was now present in the sap. On 9th May, when activity was estimated to be reaching its peak, sap was collected from the eight dis-budded branches and from eight control branches on the adjacent trees. The sap was concentrated, chromatographed in water and bioassayed by the techniques already described. The results presented in Figs. 1 and 2 show an essentially similar pattern of hormone activity in branches with and without buds. Quantitative assay showed no significant differences in cytokinin activity or in mesocotyl activity as a result of dis-budding (Table I). It is concluded from this experiment that the hormones in the xylem sap of the apple do not have their origin in the buds.

Table I

Effect of dis-budding branches on 1st April on the hormone content of the xylem sap of apple 'Stoke Red' on Crab C rootstock

Date 1966	Stage of development of buds	Buds removed or attached	Substance K Soybean callus assay µg equivs. kinetin/l.	Substance A Oat mesocotyl assay µg equivs. IAA/l
April 1	Dormant	Attached	0	0·7
April 21	Swelling	Attached	45	6
May 9	Breaking	Attached	91	56
May 9	Breaking	Removed	125	49

Discussion

One possible criticism of these experiments is concerned with the method of sap extraction, which involves continual cutting of the distal end of the branch in order to maintain the sap flow from the lower end under suction.[18] If this is not done, the tracheids quickly become blocked by air, which prevents further extraction of sap. Cutting the branch with secateurs inevitably involves some damage to the living cells of the wood and adjacent tissues, sap from which could conceivably contaminate the extracted xylem sap. Although the possibility of some contamination must be admitted, it is considered that it would not be nearly extensive enough to account for the high concentration of hormone activity in the sap collected in spring. Moreover, in preliminary work with bleeding sap of birch and other trees, collected by boring a hole in the trunk just above soil level, similar levels of hormone activity were found. In bleeding sap collected by this method, contamination by sap from living cells would be expected to be even less than in sap extracted under suction.

If the possibility of an artefact be excluded, some other explanation must be found to account for the presence of these biologically active substances in the dead tissues of the xylem. Recent work by Peel[24] has shown that ions and ^{14}C labelled assimilates in willow can move radially from the extracambial cells into the xylem tissues, and be passed into the sap stream within the xylem vessels. Moreover, this radial transport of ions into the sap stream shows a definite seasonal variation, being high in March and April, but low at other times of the year. If ions and carbohydrates can move in this way, it seems likely that hormonal substances could reach the xylem stream by a similar pathway. It is probably significant that the seasonal changes in the concentration of cytokinin and Substance A in the sap follow similar patterns to those already reported for the mineral constituents of the sap,[18] and for the amide and amino acid concentration.[25] All these substances show a rapid increase in concentration in early spring and reach their maximum concentration at about the time of full bloom. Subsequently there is a decrease in concentration, which reaches a very low value by August. A rather similar pattern has been found for the concentration of reducing sugars in the tracheal sap of pear and apricot trees.[26] The actual concentration of these materials in the xylem sap might be expected to depend, not only on the rate of efflux into the sap stream, but also on the rate of water uptake. It has already been suggested that the temporary drop in concentration of Substances K and A in 'Cheddar Cross' sap at the end of April might be a dilution effect following the exceptionally high rainfall that month.

Whatever the precise pathway by which these hormonal substances enter the xylem it is clear that once they are in the sap stream, the way is open for transport to all the aerial portions of the tree. The rate at which these substances can be moved to the leaves, fruits and apical growing points will depend not only on the concentration in the sap, but also on the rate of ascent of the sap. In the early spring, before the leaves are expanded, there is a movement of sap from roots to shoots along a tension gradient, but this movement is slow compared with that in a transpiring tree in full leaf. Huber & Planki[27] using a thermoelectric method to study sap flow in intact trees of *Fagus sylvatica*, found that in March and April the rate of ascent of sap was between 0·1 and 0·2 m/h, rising to over 1·0 m/h in June, when the tree was fully leafed out.

Although it is premature to assess the physiological significance of these hormones in the xylem sap, a number of interesting possibilities present themselves. The appearance of cytokinins in the sap at the time of bud break in spring, coupled with the observations of Chovjka *et al.*[28] that dormant buds can be stimulated into growth by injecting them with cytokinins, suggests a causal relationship between these two phenomena. Such a conclusion, however, would be at variance with pomological experience, which shows that it is the character of the scion variety, not the rootstock, which determines the times of bud-break. A graft of a late-leafing variety, top-worked on a tree of an early variety, will retain its late-leafing habit, which suggests that the signal for bud-break comes from the buds themselves rather than from the rootstock.

A rather better case can be made for the relation between cytokinins in the xylem sap and extension growth of shoots. It was observed that in the 'Stoke Red' apple seedlings, shoot growth had ceased by the end of July, whereas in the much more vigorous 'Cheddar Cross' growth continued well into September. Thus, in both these groups of trees the cessation of extension growth corresponded approximately to the time when Substances K and A disappeared from the xylem sap. Internode extension in the growing shoot is believed to be controlled by gibberellin from the young leaves, possibly in conjunction with auxin from the stem apex, but work on the sterile culture of apple shoot tips has shown that the maintenance of meristematic growth in both these regions requires a source of cytokinin. In growing shoots it seems probable that this cytokinin is supplied from the roots via the xylem sap. Such a conclusion would be consistent with the fact that, in shoots separated from their root systems, meristematic activity quickly ceases and is only renewed if adventitious roots develop. Mullins[29] has shown that in cuttings of *Vitis vinifera* removal of adventitious roots causes the young inflorescences to atrophy and that this morphogenetic effect of the roots can be replaced by synthetic cytokinins, such as benzyl adenine, applied either to the base of the cutting or directly to the emergent inflorescences. Cytokinins are known to play a role in leaf growth, in fruit-set and in the control of senescence and it seems probable that all of these processes could be influenced by the cytokinins carried in the xylem sap stream.

The presence of the unknown growth-promoting substance A in the xylem sap is of particular interest. Its concentration in the spring was such that only 0·3 ml of undiluted sap was required to give a strong reaction in the oat mesocotyl test. Nitsch[9] has reported the presence of inactive auxin precursors in the sap of sugar maple and vine, which give rise to active auxins on autoclaving, but it seems unlikely that Substance A is of this nature, since the activity was detected and measured in un-autoclaved sap. The extremely polar nature of Substance A suggests that the active material may be linked with a sugar molecule or other water soluble moiety, and the separation into two active fractions by electrophoresis could indicate that two different substances are involved. The higher concentrations of Substances A and K in the sap of the seedling trees studied in 1964–65 as compared with the mature grafted trees studied in 1966 should be noted. This could well be a seasonal or varietal difference, although the possibility of a connexion between xylem sap hormones and juvenility would seem to warrant further investigation. It would also be interesting to compare, both qualitatively and quantitatively, the hormone content of the xylem sap from vigorous apple rootstocks such as M.XXV or M.XVI and from a rootstock such as M.IX which dwarfs the scion varieties grafted upon it and which induces a relatively early cessation of extension growth.

The presence in the ascending sap stream of these biologically active compounds raises the interesting problem of their ultimate fate when they reach the leaves, fruits or other aerial organs of the tree. Certain organs, particularly the seeds, are known to accumulate relatively high concentrations of cytokinins and gibberellins at particular stages of development, but there is, as yet, no evidence that these hormones enter the seed in the sap stream. Indeed, what evidence there is points rather to biosynthesis *in situ*. If these sap hormones are not accumulated at their destination then it must be supposed, either that they are metabolised to nonactive compounds, or that they are re-exported in the phloem. Although much remains to be discovered, it is clear that the recognition of the xylem as a major pathway of hormone translocation must profoundly affect opinions on problems concerned with the hormonal control of plant growth and development.

References

[1] Went, F. W., *Pl. Physiol., Lancaster*, 1943, **18,** 51
[2] Chibnall, A. C., & Wiltshire, G. H., *New Phytol.*, 1954, **53,** 38
[3] Richmond, A., & Lang, A., *Science, N.Y.*, 1957, **125,** 650
[4] Kulaeva, O. N., *Fiziol. Rast.*, 1962, **9,** 229
[5] Loeffler, J. E., & Overbeek, J. van, *Colloq. Int. Centre Nat. Rech. Sci., Paris*, 1964, **123,** 77
[6] Kende, H., *Science, N.Y.*, 1964, **145,** 1066
[7] Miller, C. O., 'Modern methods of plant analysis', 1963, **6,** 194 (Ed. Linkskens, H. F., & Tracey, M. V.) (Berlin: Springer Verlag)
[8] Itai, C., & Vaadia, Y., *Physiologia Pl.*, 1965, **18,** 941
[9] Nitsch, J. P., & Nitsch, C., *Bull. Soc. bot. Fr.*, 1965, **112,** 11
[10] Seth, A., & Wareing, P. F., *Life Sci.*, 1965, **4,** 2275
[11] Weiss, C., & Vaadia, Y., *Life Sci.*, 1965, **4,** 1323
[12] Carr, D. J., & Burrows, W. J., *Life Sci.*, 1966, **5,** 2061
[13] Skene, K. G. M., & Kerridge, G. H., *Pl. Physiol., Lancaster*, 1967, **42,** 1131
[14] Jones, O. P., *A. Rep. East Malling Res. Stn.* 1964, 1965, p. 119
[15] Carr, D. J., Reid, D. M., & Skene, K. G. M., *Planta*, 1964, **63,** 382
[16] Phillips, D. J., & Jones, R. L., *Planta*, 1964, **63,** 269
[17] Skene, K. G. M., *Planta*, 1967, **74,** 250
[18] Bollard, E. G., *J. exp. Bot.*, 1967, **4,** 363
[19] Frankland, B., & Wareing, P. F., *Nature, Lond.*, 1960, **185,** 255
[20] Nicholls, P. B., & Paleg, L. G., *Nature, Lond.*, 1963, **199,** 823
[21] Whyte, P., & Luckwill, L. C., *Nature, Lond.*, 1966, **210,** 1360
[22] Luckwill, L. C., *J. hort. Sci.*, 1956, **31,** 89
[23] Nitsch, J. P., & Nitsch, C., *Pl. Physiol., Lancaster*, 1956, **31,** 94
[24] Peel, A. J., *J. exp. Bot.*, 1967, **18,** 600
[25] Tromp, J., & Ovaa, J. C., *Z. PfiPhysiol.*, 1967, **57,** 11

[26] Andersen, F. G., *Pl. Physiol., Lancaster*, 1929, **4,** 459
[27] Huber, B., & Planki, L., *Forstwiss. Zbl.*, 1956, **75,** 350
[28] Chovjka, L., Veresz, K., & Kozel, J., *Biologia Pl.*, 1961, **3,** 140
[29] Mullins, M. G., *J. exp. Bot.*, 1967, **18,** 206

INDOLES OF MAIZE AND THE SOURCE OF WOAD INDIGO

By B. B. STOWE, E. EPSTEIN and M. VENDRELL*

(*Kline Biology Tower, Yale University, New Haven, Connecticut, U.S.A.*)

Two standard plant sources of indole derivatives have been examined by improved techniques of gradient elution, thin-layer, and vapour-phase chromatography. In sweet corn varieties of maize, large differences were found between varieties, and conventional extraction procedures produced artefacts. Cold alcohol failed to completely extract indole derivatives, while boiling alcohol produced large quantities of the indoleacetyl ester of the alcohol used, suggesting that transesterification of the natural indoleacetyl esters occurs. Brief exposure to ammonium hydroxide produced indoleacetamide, evidently by ammonolysis. Even pH 7·4 buffer released indoleacetic acid from the ester complex. Extraction at −78° with methylene chloride followed by isopropanol minimises these difficulties, but requires an acetonitrile–hexane partitioning to remove lipids. About 1 mg/kg of indole-3-acetic acid and indole-3-carboxaldehyde were found; less indole-3-carboxylic acid was present.

Woad (*Isatis tinctoria* L.) contains an indigo precursor which is *not* indican as commonly supposed, but the material named isatan by Beijerinck in 1901.[20] This substance was renamed isatan B, purified, and from its alkaline hydrolysate the crystalline calcium salt of 5-ketogluconic acid was obtained, and further characterised as the semicarbazide. Isatan B is thus indoxyl-5-ketogluconate. This is the first biological occurrence of 5-ketogluconic acid above the bacteria; it is a plausible precursor of ascorbic acid. Woad also contains glucobrassicin and neo-glucobrassicin, and a similar indole compound not yet identified.

Introduction

Conclusive chemical identifications of indoles from plant materials have for the most part been limited to those from tissues in which they occur in relatively large quantities. Examination of the indole substances naturally active as auxin hormones in plants has been much more difficult, and for the most part has relied upon bio-assay. New sensitive techniques of chemical analysis, notably gas chromatography, theoretically approach the limits necessary for identification of hormones in the quantities in which they normally occur, but in practice they are exceedingly difficult to apply with unequivocal results. In the present work two relatively rich sources of indole derivatives have been examined in the hope that deficiencies in the new methods would be un-

*Present address: Division of Food Preservation, C.S.I.R.O., P.O. Box 43, Ryde, N.S.W., Australia.

covered before taking up the more difficult task of studying typical plant materials.

Experimental

Various chromatographic techniques pioneered in this laboratory were employed. The gradient elution column chromatography was essentially that of Powell,[1] as modified by Grunwald *et al.*,[2] while thin-layer solvents of Perley & Stowe[3] were also utilized. Maize kernels were of the sweet corn variety, 'Country Gentleman', one of the classic sources of indoleacetic acid in plants.[4,5] They were chilled to dry ice temperatures when they could be easily ground to 80 mesh in a Wiley mill. The woad plants were grown in Yale University's Marsh Botanic Garden from seed obtained from the herb garden of Sissinghurst Castle, Kent, or from Thompson & Morgan (Ipswich) Ltd. Seedling and mature plant material was used at different times of the year, and did not show any marked variation in the nature and content of the indole compounds present.

Indoles in maize

Maize meal was extracted by cold ethanol at —10°, the extract evaporated, then partitioned into acidic, basic, neutral, and water-soluble fractions,[1] and applied to gradient elution columns. Fractions from the column were then applied to thin-layer plates which were developed in 2-butanone/hexane[2] and sprayed with Ehrlich reagent. With the glutinous maize endosperm, cold alcohol proved unsatisfactory, and consistent results were not obtained, apparently because of incomplete extraction.

Accordingly, boiling ethanol was used as the extracting agent. Such extracts led to large spots on the thin-layer plates, which had the gradient elution, thin-layer, and gas chromatographic characteristics of ethyl indoleacetate. This compound has been reported from maize before, but later was shown to be an artefact.[6] When boiling methanol was used as extractant, again a large spot was found in the neutral fraction, but this time with every chromatographic characteristic of methyl indoleacetate. It would appear that these substances were formed during the boiling in alcohol; in this medium they are unlikely to arise by esterase action. Since maize kernels contain large quantities of natural indoleacetyl esters, it is most likely that transesterification has occurred, as is frequently the case with fatty acid esters upon refluxing in alcohol.

In some neutral extracts prepared in this way another prominent compound was found which had the chromatographic behaviour shown in the three systems of indole-3-acetamide. Its presence was observed when ammonia had been used to change the pH of the solution during partitioning. Although exposure to the ammoniacal mixture had been brief, it was sufficient to allow ammonolysis of the natural indoleacetyl esters, as has been reported by Zenk.[7]

The formation of indoleacetic acid from a high molecular weight auxin precursor by alkaline treatment of this variety of maize was reported by Berger & Avery.[4] Since these investigators largely used high pHs and high temperatures, the finding that even pH 7·4 buffer led to the production of 3-indoleacetic acid from a maize extract after standing for a few hours at room temperature was not expected, although such a claim was made by Stehsel.[8] The precursor of the free acid seems to be a high molecular weight material as it was only slightly soluble in all solvents tested. Its low solubility in both polar and non-polar solvents seems to exclude its being identical to the indoleacetyl inositols reported by Nicholls,[9] the zeanins reported by Farrar,[10] or to the indoleacetyl glucosides reported by Sahai Srivastava[11] and Zenk.[7] However, it could easily be a higher polymer of one of these, but as a means of purifying it has not yet been developed, because of its low solubility, the substances cannot immediately be compared. That several substances have been reported is likely to be due to the different varieties of maize employed. It has been found that maize kernels from different varieties are strikingly different in their hormone constitution. (see also Teas & Newton,[12] and Tafuri[13]).

Since the conventional extraction methods had not succeeded in removing the natural indole derivatives unchanged, a new extraction scheme was devised.[14] In this the 500 g of maize was dropped into a 1 l methylene chloride bath at dry ice temperatures, as soon as it was ground. The mixture was stirred from 4 to 6 h, keeping the temperature at —78° with dry ice. After the methylene chloride had been filtered off 1 l of isopropanol was added, also at dry ice temperatures, for a second period of extraction of 4 to 6 h. After filtration, both of the organic extracts were evaporated *in vacuo* at temperatures not in excess of 50°. The resulting combined residue, weighing 36 g, was largely lipoidal in nature, and to remove these lipids the residue was partitioned between acetonitrile and hexane. This method efficiently transferred the majority of the lipids to the hexane layer from which nearly 20 g were recovered. A large solid gel formed at the interface and was not readily dissolved; but since thin-layer chromatograms showed this not to contain significant amounts of indoles, it was ignored in later work. The hexane layer does have non-Ehrlich reacting materials which liberate indoles on alkaline hydrolysis, and should be worthy of further study. The simple indole derivatives were found in the acetonitrile layer from which 300 mg of solids were obtained after evaporation. This method therefore achieves a significant purification as well as a more complete extraction of simple indoles from the difficult maize kernel tissue.

The acetonitrile residue was partitioned between methylene chloride and sodium bicarbonate.[14] From this 10 mg of acidic compounds were obtained and 206 mg of neutral compounds, while no Ehrlich-reactives were noted in the basic and water-soluble fractions. The first two fractions were then submitted to the combination gradient elution and thin-layer methods described earlier. In the acid fraction a prominent spot corresponding in every chromatographic characteristic to indoleacetic acid was obtained, and after diazomethylation this behaved as the methyl ester of indoleacetate on gas chromatography. Approximately 1 mg/kg of the original kernels was found. In the same acid fraction about one-hundredth of this quantity of a material corresponding to indole-3-carboxylic acid was found. The neutral fraction contained several substances reacting to 2,4-dinitrophenylhydrazine. One of these had the characteristics of indole-3-carboxaldehyde, and its presence was substantiated by the three-peaked ultra-violet absorption spectrum which is characteristic of this compound.[5] Its concentration was about the same as that of indole-3-acetic acid.

A large number of other Ehrlich-reacting spots were found in both the acid and neutral fractions, but again, at least some of these were artefacts. Prominent among these were dark green Ehrlich-reactive compounds occurring in nearly every fraction of the gradient elution and which migrated to R_f 0·7 to 0·8 on the thin-layer plate developed in 2-butanone/hexane (1:3). These could not be identified, but did not appear in the original extract; apparently they arise as a result of oxidation on the gradient elution column. The substance from which they are produced does not give Ehrlich, Salkowski or Procházka reactions, nor was indolic ultra-violet absorption observed, and accordingly there is no certainty that an indole precursor is, in fact, involved. Other substances present cannot yet be conclusively identified.

The major conclusion of this work is that several simple laboratory procedures, normally considered to be unobjectionable, produce changes in extracts from maize. As a consequence care should be taken before even prominent substances in these extracts are reported as being present in the native material. Moreover, as dry seeds are not growing tissue, the physiological significance of any substances found in them must remain in doubt.

Indoles in woad

Since normal plant tissues contain only small amounts of indole derivatives, the enzymes synthesising these are not likely to be present in substantial amounts, and indeed the history of their study has not been a very successful one.[5,15] As some plants contain large quantities of indole derivatives, these should be the most promising sources of enzymes of indolic synthesis. The maize kernel material used above is laid down during only a relatively short period of the plant's life, and it was therefore desirable to use plant tissues in which the time factor might be less critical.

As a result, attention was given to the numerous species of the genus *Indigofera* which have been agricultural sources of the dye indigo since ancient times. In some of these the indigo precursor has been identified as 'plant indican,' chemically indoxyl-β-D-glucoside, and it may reach a concentration of several % dry weight. Attempts to grow any *Indigofera* species in sufficient quantities for it to be a convenient laboratory material were unsuccessful. Accordingly, attention was turned to woad, *Isatis tinctoria* L., which was extensively cultivated in temperate climates as a source of indigo, and in which concentrations of up to several tenths % dry weight of the indigo precursor are attained.[16,17,18] Woad was easily grown in quantity in the field or in the glass house and all parts of the plant have been found to contain the indigo precursor.

However, examination of the literature disclosed a peculiar anomaly whose history is described in more detail elsewhere.[19] Briefly, most reference works state that the indigo precursor in woad is plant indican, namely indoxyl-β-D-glucoside. However, two papers by Beijerinck,[20] which appear to have been virtually ignored, present clear evidence that the indigo precursor in woad cannot be a glycoside since it is hydrolysed by weak bases and does not produce indoxyl in the presence of acids. Moreover, enzymes which hydrolyse the woad precursor do not act on indican and vice versa. Beijerinck named the material in woad 'isatan', not realising this name had been used for another substance previously. Accordingly, it is proposed to rename it 'isatan B'.

The fundamental reason for the perpetuation of the error that indican is present in woad can be traced to the fact that Schunk originally created the name indican for the material in woad.[21]

However, he, and many others, later loosely applied the name 'indican' to indigo precursors in other plants.[22] After the identification of the indican of *Indigofera* species and *Polygonum tinctorium* Ait. as indoxyl-β-D-glucoside,[23] it was presumed that the precursor in woad must be the same. It is noteworthy that Schunk himself later realised this was an error and stated his view that the two materials must be different.[24] In particular, he noted that the 'peculiar sugar'[21] of woad was not fermentable and could not be glucose. His remarks appear to have been completely ignored and as a result, isatan B has escaped examination for over 60 years.

The correctness of Beijerinck's views was established in thes laboratories by Nabors (personal communication), using thin-layer and column chromatographic techniques. A blue spot developing after spraying with mild bases was found at R_fs completely different from those of indoxyl-β-D-glucoside. The latter compound has not been found in any woad extract.

The first attempts to isolate isatan B by these techniques did not succeed because of the relatively low yields obtained on extraction, and here too Beijerinck's observations are pertinent. He noted that the pH of the extraction medium was critical and must neither be too acidic nor approach neutrality and recommended the use of an oxalate buffer. In the present work 0·2 M sodium formate buffer at pH 3·5 has been used with greater success. Plunging woad leaves into boiling buffer is effective, but 50° suffices to destroy the hydrolysing enzyme 'isatase' and generally leads to the highest yields of isatan B. However, it may also be extracted from quick-frozen lyophilised woad leaves by dry ethyl acetate and therefore it is present naturally and is not an enzymatic artefact introduced under aqueous conditions.

Isatan B is soluble in water, alcohols and ethyl acetate. It can be transferred to the latter solvent from water by several partitionings. Evaporation of the dried ethyl acetate produces a residue which can be fractionated on a silica gel column prepared with chloroform and eluted with increasing quantities of ethyl acetate. The presence of isatan B was detected by formation of indigo when weak alkali was added in the presence of air, and could be quantitatively followed with indigo sulphonation analysis or by colorimetry of its dimethyl sulphoxide solution. These assays disclosed that isatan B is readily destroyed, indirubin being produced, in acidic solutions, while the formation of other unknown pigments was also observed often. All attempts to prepare isatan B, or its trinitrobenzene adduct, in a crystalline form have failed, but the quantitative indigo assays indicated that assuming its structure to be as indicated below, some preparations were 30% pure.

In the products of the hydrolysis of these semi-purified preparations, the presence of indoxyl was confirmed by the fluorindal test and ultra-violet absorption, as well as by its subsequent oxidation to a blue substance spectrophotometrically indistinguishable from authentic indigo. Since its solubility properties led to the belief that isatan B might be an indoxyl ester of a sugar acid, paper chromatograms of its alkaline hydrolysate were developed with several carbohydrate solvents. These results have been summarised elsewhere.[19] Briefly, they indicated that a single substance with typical carbohydrate reactions was present which was distinctively different from all common sugar acids. The most striking characteristic of this compound was its ability to reduce 2,6-dichlorophenol indophenol, indicating a dienol type structure. Further paper chromatography led to the identification of this sugar acid as 5-ketogluconic acid. Identification was confirmed by isolation of the calcium salt and the semicarbazide, both of which were identical with synthetic samples. Quantitative assays of the amount of this carbohydrate indicated a one to one correspondence with the amount of indoxyl present. Since no other carbohydrate could be detected, the structure of isatan B must be indoxyl-5-ketogluconate in the furanose form since isatan B itself does not reduce 2,6-dichlorophenol indophenol.

This natural product has not been described before, and indeed 5-ketogluconic acid is only previously known from *Acetobacter*. However, it is of exceptional interest in that it is a logical precursor of L-ascorbic acid whose biogenesis in plants is still uncertain.[25] It should be emphasised that even though isatan B is an indole compound releasing an ascorbic acid-like substance on hydrolysis, it cannot be directly compared with ascorbigen, a substance isolated from several other crucifers. This is now known to be an artefact formed from glucobrassicin in this tissue, whose decomposition products condense with ascorbic acid.[26,27] The structure of ascorbigen has recently been suggested by Kiss & Neukom to be an indole-3-methylene linked to C-2 of ascorbic acid.[28] It has also been found that this material readily appears in woad extracts, where it may also arise from glucobrassicin which is present in large quantities. Elliott (personal communication) has also found neoglucobrassicin and another glycosinolate indole of unknown structure to be present.

Woad, therefore, promises to be material of special suitability for the study of indoles, sulphur compounds and carbohydrate metabolism, and should rapidly regain interest.

Acknowledgments

This research was sponsored by grants from the U.S. Public Health Service (GM-06921) and the Whitehall Foundation. The authors are grateful to Drs. M. Kutáček and Z. Procházka for helpful suggestions.

References

[1] Powell, L. E., *Pl. Physiol., Lancaster*, 1964, **39,** 836
[2] Grunwald, G., Vendrell, M., & Stowe, B. B., *Analyt. Biochem.*, 1967, **20,** 484
[3] Perley, J. P., & Stowe, B. B., *Biochem. J.*, 1966, **100,** 169
[4] Berger, J., & Avery, G. S., jun., *Am. J. Bot.*, 1944, **31,** 199, 203
[5] Stowe, B. B., *Fortschr. Chem. org. NatStoffe*, 1959, **17,** 248
[6] Fukui, H. N., *et al.*, *Nature, Lond.*, 1957, **180,** 1205
[7] Zenk, M. H., *Nature, Lond.*, 1961, **191,** 493
[8] Stehsel, M. L., Thesis Univ. California, 1950
[9] Nicholls, P. B., *Planta*, 1967, **72,** 258
[10] Farrar, K. R., *Biochem. J.*, 1962, **83,** 220
[11] Sahai Srivastava, B. I., *Pl. Physiol., Lancaster*, 1963, **38,** 473
[12] Teas, H. J., & Newton, A. C., *Pl. Physiol., Lancaster*, 1951, **26,** 494
[13] Tafuri, F., *Phytochem.*, 1966, **5,** 999
[14] Stowe, B. B., Vendrell, M., & Epstein, E., 'Physiology and biochemistry of plant growth regulators', 1968 (Eds. Setterfield, G., & Wightman, F.) (Ottawa: Runge Press)
[15] Mahadevan, S., 'Modern methods of plant analysis', 1964, **VII,** p. 238 (Eds. Paech, K., & Tracey, M. V.) (Berlin: Springer)
[16] Planer, J. J., *Akademie gemeimuitzger Wiss. zu Erfurt Acta*, 1779/80, 34
[17] Giobert, 'Traité sur le Pastel', 1813 (Paris: Imprimerie Impériale)
[18] Hurry, J. B., 'The woad plant and its dye', 1930 (London: Oxford University Press)
[19] Epstein, E., Nabors, M. W., & Stowe, B. B., *Nature, Lond.*, 1967, **216,** 547
[20] Beijerinck, M. W., *Proc. K. ned. Akad. Wet.*, 1900, **2,** 120; *ibid.*, 1901, **3,** 101
[21] Schunck, E., *Phil. Mag.*, 1855, **10,** 73, *ibid.*, 1858, **15,** 29, 117, 183
[22] Schunck, E., *Chem. News, Lond.*, 1879, **39,** 119, 129, 143
[23] Hoogewerf, S., & ter Meulen, H., *Proc. K. ned. Akad. Wet.*, 1900, **2,** 520
[24] Schunck, E., *Chem. News, Lond.*, 1900, **82,** 176
[25] Mapson, L. W., 'The vitamins', 1967, **I,** p. 369 (Eds. Sebrell, W. H., & Harris, R. S.,) (London: Academic Press)
[26] Gmelin, R., & Virtanen, A. I., *Suomal. Tiedeakat. Toim.*, 1961, **AII,** No. 107
[27] Kutáček, M., & Procházka, Ž., 'Régulateurs naturels de la croissance végétale,' 1964, p. 445 (Paris: Centre National de la Recherche Scientifique)
[28] Kiss, G., & Neukom, H., *Helv. chim. Acta*, 1966, **49,** 989

DISCUSSION

Dr. B. J. Heywood.—Is it necessary to invoke transesterification for the formation of the supposed artefact IAA methyl and ethyl esters? In boiling alcohol one would expect some direct esterification to occur.

Professor Stowe.—The hazard of transesterification is becoming increasingly appreciated by lipid biochemists, and must be considered whenever alcohols are used with esters. In this case, the quantities of IAA ester which appear seem larger than the free IAA present. Moreover, in our experience, some catalyst like BF_3 is required for ready esterification of free IAA. It is noteworthy that Nicholls, also working with maize, has reported[9] indoleacetyl migration on inositol, an analogous process.

NATURAL CYTOKININS

By J. P. NITSCH

(*Laboratoire de Physiologie Pluricellulaire, Le Phytotron, C.N.R.S., Gif-sur-Yvette, France*)

Natural cytokinins are essentially adenines substituted in the N_6 position. They may occur as free bases or in the form of nucleosides and nucleotides. Methods for their extraction, purification and assay are reviewed briefly. Cytokinins circulate in the sap probably in the form of nucleotides and exert a wide range of biological effects which can be grouped in three main categories, namely, (1) effects not linked to cell division and consisting essentially in the promotion of protein synthesis, (2) effects linked to cell division and in which a particular type of DNA seems to be involved, and (3) effects linked to the differentiation of buds, in which adenine has a prominent role.

Introduction

As defined by Skoog *et al.*,[1] cytokinins are substances which are necessary for the process of cytokinesis, i.e. the division of a cell into two daughter cells.

The study of cytokinins goes back to the pioneering work in 1941 and 1942 of van Overbeek *et al.*[2,3] who used 'coconut milk' (that is the liquid endosperm of *Cocos nucifera*) to stimulate the growth of immature plant embryos. In 1948, Caplin & Steward[4] showed that coconut milk stimulated vigorous callus proliferation in another system,

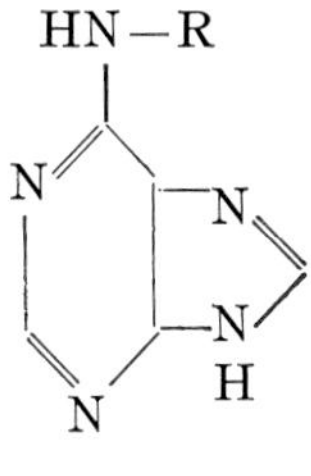

Adenine: R=H

Zeatin: R=

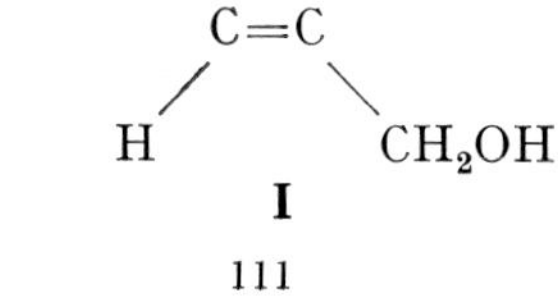

I

$$\text{Dihydrozeatin: R} = -CH_2-CH_2-CH\begin{array}{l}\diagup CH_3\\ \diagdown CH_2OH\end{array}$$

II

$$\text{Isopentenyladenine: R} = -CH_2-CH{=}C\begin{array}{l}\diagup CH_3\\ \diagdown CH_3\end{array}$$

III

$$N_6\text{-Methyladenine: R} = -CH_3$$

IV

the culture of carrot root tissues. After much effort by different laboratories, it became apparent that coconut milk owed its biological activity to a combination of different factors, namely: (a) auxin, which, as indicated by van Overbeek[3] and shown by Nitsch & Nitsch,[5] increased after autoclaving, probably because of the hydrolysis of some bound form; (b) reduced nitrogen;[6] (c) a neutral fraction, composed of sorbitol (15 g/l), *scyllo*-inositol (500 mg/l) and, especially, *myo*-inositol (100 mg/l);[7] (d) cytokinins[8,9] which were isolated by Letham in 1968[10] in the form of the riboside of zeatin.

Historically, however, the first natural cytokinin was in fact isolated not from coconut milk but from immature kernels of maize by Letham in 1963.[11] It was identified as a 6-substituted adenine with a 4-hydroxy-3-methylbut-*trans*-2-enyl side-chain (**I**). After this breakthrough, several other natural cytokins have been isolated.

Tests for cytokinins

As usual, one of the crucial factors for success is devising a suitable test. Since most, if not all, of the natural cytokinins are substituted purines, it might be supposed that their absorption in the u.v. range could be used as a guide through the various purification steps. However, many other compounds absorbing in the same region exist in plant extracts, including adenine itself and its nucleoside and nucleotides.

Another difficulty is that cytokinins are present in minute

amounts, the highest yield being 0·7 mg in 60 kg of immature maize kernels.[12] Biological tests generally are superior to other tests in being more sensitive and more specific. In the case of cytokinins, the obvious candidates are tests involving the proliferation of cells.

Perhaps the most specific and better-known test involves the use of calibrated pith tissue explants of the tobacco stem, c.v. 'Wisconsin 38'. In this laboratory other varieties have also been tested, such as 'Maryland Mammoth' kept in a vegetative state by long photoperiods. 'Maryland Mammoth' was just as specific as 'Wisconsin 38'. On the other hand, *Nicotiana glauca*, although responding vigorously to the combination of IAA+kinetin, does grow to some extent with either IAA or kinetin alone. In this test, the fresh and dry weights of the cultures are determined after 21 days of culture.

Although qualitatively very reliable, the tobacco pith test does not give good quantitative results, as there is only a small portion of the curve over which a linear relationship exists between cytokinin concentration and growth increment. A second test of this kind involves the use of soybean callus. In this case, according to Miller,[13] the range of proportionality is much greater than with tobacco pith.

Tests which require much less time (1–2 days) involve the growth of leaf discs[14] or of fronds of Lemnaceae such as *Lemna minor*[15] or *Spirodela oligorrhiza*.[16] The second type of test is not very specific, as adenosine and cobaltous ions also stimulate growth.

Other biological assays are based upon the ability of cytokinins to delay the breakdown of chloroplasts. Such methods measure the chlorophyll content of detached leaves after a relatively short period (2 days), for example the method of Osborne & McCalla,[17] or the more practical one of Kende[18,19] with barley leaves.

The property of cytokinins to stimulate bud formation has also been used to detect the presence of these substances, especially in the case of mosses.[20]

Several other tests are listed in two recent papers by Letham.[16,21] One which is not included, however, involves the synthesis of betacyanins in the hypocotyl of *Amaranthus*.[22–24] This test has been refined recently by Bigot[25] in this laboratory and found to be highly specific for cytokinins, adenine and methyladenine having no effect at concentrations up to 10^{-5}M, whereas all the true cytokinins tested were active in the 3×10^{-7}M to 10^{-5}M range (Fig. 1).

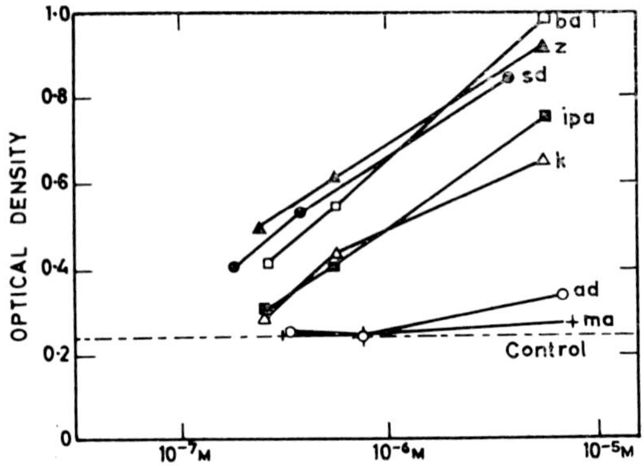

FIG. 1. *Effect of various cytokinins and related compounds on the production of betacyanins in the hypocotyl of* Amaranthus caudatus
Measured at 542 nm with a Coleman colorimeter
ba = benzyladenine, z = zeatin, sd = SD-8339, ipa = isopentenyladenine, k = kinetin, ad = adenine, ma = methyladenine. (From Bigot, 1968).[25]

Methods of extraction and purification

The chief extractants used have been alcohols (methanol or ethanol) or water. In the latter case, the tissues may be dropped into boiling water or autoclaved for a short time (5–10 min) in order to stop enzymatic activity.

Purification from the crude extracts can be achieved by various ways: (a) the use of non miscible solvents such as n-butanol and water (s-butanol is preferable as it is more easily evaporated off than n-butanol); (b) precipitation by heavy metals such as lead, mercury or silver—it is necessary to completely remove these ions before biological testing, as they are toxic to cells; (c) adsorption through ion-exchange resins, such as Dowex 50; (d) adsorption by activated charcoal in an aqueous medium, followed by elution with methanol, ammoniacal methanol or, even better, diethylamine;[26] (e) paper and thin-layer chromatography—the R_f's of a number of natural cytokinins are listed in Table I; (f) formation of a crystalline picrate.[27]

Table I

Chromatography of natural cytokinins
Thin-layer chromatography of 2 μg of cytokinins on cellulose-coated films ('MN-Polygram, Cel 300') pre-washed with the solvent

Compound	R_f in:		
	solvent 1	solvent 2	solvent 3
Isopentenyladenine	0·28	0·90	0·32
Zeatin	0·37	0·80	0·43
Dihydrozeatin	0·48	0·83	0·50
IPA riboside	0·50	—	0·76
Zeatin riboside	0·65	—	0·87
Grapevine sap: (a)	0·70	0·00	—
(b)	0·85	0·00	0·78

Solvent 1: water
Solvent 2: butanol (86%), 28% ammonia (5%), water (9%)
Solvent 3: boric acid 0·03M, pH adjusted to 8·4 with NaOH

Identification of the natural cytokinins

Natural cytokinins from plant sources

Four natural cytokinins have been isolated from plant sources: zeatin, dihydrozeatin, isopentenyladenine and methyladenine. A fifth cytokinin has been partially characterised. A possible precursor, triacanthine, has been extracted in large quantities from certain woody species of plants.

Zeatin

The first natural cytokinin was isolated from immature maize (*Zea mays*) kernels by Letham[11,12] and named zeatin for that reason. Chemically, it consists of a molecule of adenine in which one of the hydrogens of the amino group has been substituted by a 5-carbon side-chain, the 4-hydroxy-3-methylbutenyl group (**I**).

Dihydroxyzeatin

From immature seeds of yellow lupin, Koshimizu *et al.*[28] isolated an active cytokinin which differs from zeatin in that its side-chain consists of the 4-hydroxy-3-methyl-butyl group (**II**).

Isopentenyladenine

From cultures of *Corynebacterium fascians*, an organism which produces fasciation and adventitious bud production in certain trees, a third cytokinin has been extracted which is also a relative of zeatin, namely isopentenyladenine (**III**).[29] In the form of its nucleoside, isopentenyladenine occurs in the t-RNA of yeast,[30,31] and of mammalian tissue.[32] Hall *et al.*[33] have shown that isopentenyladenosine occurs also in plant t-RNA, together with the *cis* isomer of the zeatin riboside. In maize, they found only the zeatin riboside in t-RNA (no isopentenyl riboside).

Methyladenine

6*N*-Methyladenine (**IV**) has been isolated from cultures of *Corynebacterium fascians* by Helgeson & Leonard[29] and found to possess a small cytokinin activity.[34,35]

Crown-gall substances

Wood[36] has extracted from crown-gall tissues of *Vinca rosea* two cytokinins which seem to be nicotinamide hexose compounds containing a sulphur atom and a methyl group. The exact identity of such non-purine cytokinins is still unknown.

Triacanthine

An isomer of isopentenyladenine, triacanthine, is very abundant in young leaves of *Gleditsia triacanthos* L., in *Holarrhena floribunda* and *Chidlovia sanguinea*.[37] Leonard & Deyrup[38] had decided that the compound was 7-isopentenyladenine, but Cave *et al.*[37] showed that the isopentenyl chain was actually located at the 3 position. Triacanthine is, therefore, 3-(γ,γ-dimethylallylamino) purine.

Although Beauchesne & Goutarel[39] reported a small biological activity for triacanthine, it was demonstrated by Rogozinska *et al.*[40] that triacanthine is totally inactive as a cytokinin, but that autoclaving may cause a shift of the side-chain from the 3 to the 6 position. Triacanthine itself is even inactive on the growth of isolated leaf discs of *Gleditsia*.[41] At the present time, the role of triacanthine is not known and its use by the plant as a precursor for cytokinins has not been established.

Cytokinin derivatives

During fractionation of extracts of immature kernels of maize, Miller[42] was able to separate two active bands by paper chromatography in saturated s-butanol: a slow-moving band, A, and a fast-moving one, B. On rechromatography of band B in 0·03M borate buffer (pH 8·4), he obtained two active spots, B_1 and B_2.

The faster moving band, B_1, was thought to be a riboside of zeatin, while B_2 was undistinguishable from zeatin. Band A could be hydrolysed by alkaline phosphatase into B_1 and was tentatively identified as the ribotide of zeatin. These compounds were later isolated by Letham and found to be the 9-β-D-ribofuranosylzeatin-5′-phosphate[43] and the 9-β-D-ribofuranosylzeatin.[27] Thus, in a given plant extract, three forms of the same cytokinin may often coexist, namely the free base and the corresponding riboside and ribotide. The two first substances may represent degradation products of the ribotide.

Another category of compounds present in extracts of immature maize kernels has been reported by Beauchesne *et al.*[44] as being composed of an adenine moiety associated in some way with an amino acid. Although substances of this type have been purified from RNA hydrolysates by Hall,[45] the identity of the amino-acid-cytokinin complex of maize has not yet been established.

Circulation of natural cytokinins

Many observations indicate that cytokinins are synthesised in roots.[46–48] Cytokinins accumulate in immature seeds and at the upper level of chicory stem cuttings.[49] They can be found in many organs: roots and tubers, leaves, seeds and fruits.[50] Cambium, at least in the spring, is also very rich in cytokinins.[51]

Natural cytokinins seem to circulate in the sap, which has been found to contain active compounds.[9,18,26,52–54] From the bleeding sap of grapevines collected in early spring, it has been possible to purify a compound which is active on the tobacco pith bioassay, has intense u.v. absorption and contains a phosphoryl group which can be split off by alkaline phosphatase. It appears to be a nucleotide of zeatin.[55] This highly water-soluble compound may actually represent the circulating form of cytokinins, the other compounds observed by Loeffler & van Overbeek being decomposition products of the ribotide.

Physiological effects and mode of action of cytokinins

The physiological effects of cytokinins may be grouped into three main classes: (1) those which are not linked with cell division, (2) those which are linked with cell division, and (3) those which are linked with differentiation.

Effects not linked with cell division

The effects not linked with cell division, such as the prevention of yellowing of detached leaves[56–58] seem to be mostly connected with the maintenance of protein synthesis. This has been shown clearly by Mothes and his co-workers in many publications. Thus, kinetin reverses the effect of chloramphenicol, an inhibitor of protein synthesis.[59,60]

The cytokinins in their involvement in protein synthesis have been found to be part of some t-RNA's, such as one of those coding for serine in yeast[30,31] and one coding for tyrosine.[61] When labelled kinetin is fed to plant tissues, it is recovered in many fractions, especially in s-RNA.[62,63] In animals, isopentenyladenosine seems to be the main cytokinin in s-RNA.[32] In plants there is found either ribosylzeatin (the *cis*-isomer, according to Hall *et al.*[33]) as in maize, or both ribosylzeatin and isopentenyl-adenosine, as in garden peas and spinach.[33] The cytokinin plays the role of an odd base and is located next to the anticodon, as shown by Zachau *et al.*[31]

Effects linked with cell division

In 1954 Jablonski & Skoog[64] showed that, for cells of tobacco pith to divide, it is necessary to supply them with both an auxin and a cytokinin. When used alone, neither substance can elicit cell division. A further step forward in the understanding of the action of cytokinins in this system has been made recently by Nitsch & Bui[65] who found that cell division can occur when a synthetic cytokinin, such as kinetin, and an auxin are supplied in *sequence*. In this case, however, the order of application of the two substances is important: proliferation takes place when kinetin is given *before* the auxin, not when the order of application is reversed (Fig. 2).

The preparatory action of kinetin is negated by inhibitors such as Actinomycin D or 5-fluorodeoxyuridine (5-FDUR). The cultures transferred to a new medium lacking the inhibitor can be induced to grow (Table II). However, thus far, this effect of Actinomycin D could not be negated by the simultaneous presence of deoxyguanosine or guanine in the medium. The effect of 5-FDUR, on the other hand, could be prevented by the addition of thymidine.[66,67]

These results can be interpreted as indicating that cytokinins, in some way, act at the DNA level since the blocking of DNA synthesis by 5-FDUR during the pretreatment with kinetin also blocks the cytokinin effect, even though no cell divisions whatsoever take place during this pretreatment. The relation between cytokinin action and DNA has similarly been brought to light using other means by van Overbeek *et al.*[68] There seems to exist a

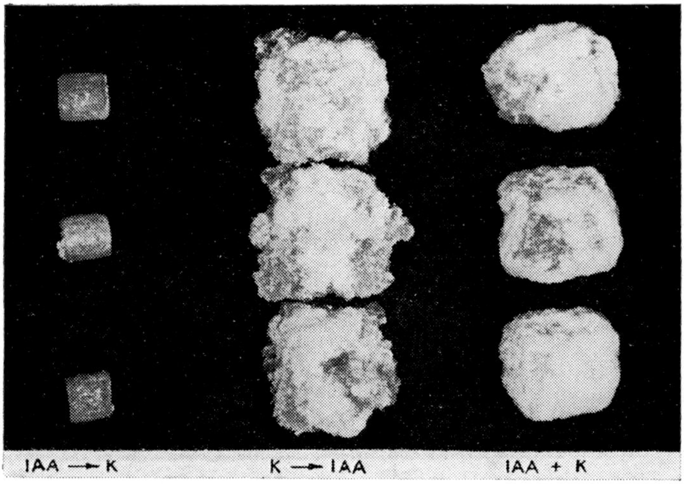

FIG. 2. *Tobacco-pith cultures grown on IAA and kinetin*

From left to right, (1) Grown for 1 day on IAA (10^{-5}M) and transferred to same basal medium with kinetin (10^{-6}M) for 20 days

(2) Grown for 1 day on kinetin (10^{-6}M) and transferred to same basal medium with IAA (10^{-5}M) for 20 days

(3) Grown on the combination of IAA (10^{-5}M) and kinetin (10^{-6}M) for 21 days

Table II

Effects of 5-*fluorodeoxyuridine and actinomycin D on the preparatory effect of kinetin*

The first treatment was given in the dark, at 25°C. Concentrations: kinetin (10^{-6}M) and IAA (1 mg/l). The 5-fluorodeoxyuridine (5-FDUR) and the Actinomycin D (Act. D) were sterilised by filtration. The weights are averages of 12 replicate cultures. Initial weights were 108 mg (fresh) and 8·6 mg (dry). (Data from Nitsch, 1967[67]).

1st treatment (5 days)	2nd treatment (16 days)	Increase in fresh wt., mg	dry wt., mg
Kinetin	IAA	190	15·3
Kin. + 5-FDUR (3×10^{-6}M)	IAA	6	1·2
Kin. + 5-FDUR + thymidine (3×10^{-5}M)	IAA	226	16·7
Kin. + 5-FDUR (3×10^{-6}M)	IAA + kinetin	−6	0
Kinetin	IAA	190	15·3
Kin. + Act. D (3×10^{-6}M)	IAA	41	4·9
Kin. + Act. D + guanine (10^{-4}M)	IAA	93	7·7
Kin. + guanine (10^{-4}M)	IAA	305	21·9
Kin. + Act. D (3×10^{-6}M)	IAA + kinetin	121	10·2
Kinetin	IAA + kinetin	242	17·3

non-chromosomal DNA fraction, which has been termed by Stern[69] 'light-DNA' and which has an important function in plant growth.

Another, rather intriguing fact has recently been brought to light, namely, that the effect of the sequential application of a cytokinin and an auxin does not occur when a natural cytokinin is used.[67] This is probably because natural cytokinins are quickly degraded by enzymes. This interpretation fits with the report that tissues contain an enzyme which can destroy IPA in 2 h, leaving some 32 other odd bases undestroyed.[71] Kinetin is known to block xanthine oxidase.[72] Perhaps synthetic cytokinins are able to block other enzymes which destroy the natural cytokinins. If this is so, the problem remains to understand why natural cytokinins, which are ineffective when administered before auxins, are very active when applied together with an auxin. Is auxin actually protecting the natural cytokinin from destruction?

Effects linked with differentiation

Cytokinins are very active in promoting the differentiation of buds in higher plants[73] and even in mosses.[74–76] It is not surprising, therefore, that a cytokinin which is very active in promoting the differentiation of buds in tobacco pith tissue has been extracted from a material which has great regenerative potentialities, namely the chicory root. From this material, Nitsch *et al.*[77] have extracted and partially purified a cytokinin which is highly water-soluble and can be precipitated by barium ions. This compound is, most probably, a nucleotide. Heide & Skoog[78] showed that other species which produce buds readily, *Begonia* and *Bryophyllum*, also contain cytokinins, one of which is probably zeatin.

The biochemical mechanism by which cytokinins promote the differentiation of buds is not known. It should be noted, however, that adenine is a powerful cofactor of cytokinins in this function, and that it is very specific, as no other purines or pyrimidines have the effect of adenine.[79]

Conclusion

At the present time, all natural cytokinins are N_6-substituted adenines, apart from the substances isolated by Wood from crown gold tissue which have not yet been unambiguously determined. Cytokinins stimulate protein synthesis and can be found in the actual molecule of certain t-RNA's. In addition, cytokinins may also play an informational role by acting on a process linked to the synthesis of DNA, possibly a 'light DNA' fraction which is especially connected with cell division.

Finally, since cytokinins are found in both plant and animal t-RNA's, it might be expected that their study will lead to discoveries pertaining to fundamental processes in living cells in general.

Acknowledgments

Grateful acknowledgment is made to Drs. D. S. Letham and G. Shaw for samples of zeatin, to Dr. X. G. Monseur for isopentenyladenine, to Dr. R. H. Hall for isopentenyladenosine, to Dr. G. Shaw for zeatin riboside and to Dr. K. Koshimizu for dihydrozeatin.

References

[1] Skoog, F., Strong, F. M., & Miller, C. O., *Science, N.Y.*, 1965, **148,** 532
[2] van Overbeek, J., Conklin, M. E., & Blakeslee, A. F., *Am. J. Bot.*, 1941, **28,** 647
[3] van Overbeek, J., *Cold Spring Harbor Symp. Quant. Biol.*, 1942, **10,** 126
[4] Caplin, S. M., & Steward, F. C., *Science, N.Y.*, 1948, **108,** 655
[5] Nitsch, J. P., & Nitsch, C., *Beitr. Biol. Pfl.*, 1955, **31,** 387
[6] Steward, F. C., & Shantz, E. M., *Année biol.*, 1954, **30,** 139
[7] Pollard, J. K., Shantz, E. M., & Steward, F. C., *Pl. Physiol., Lancaster*, 1961, **36,** 492
[8] Kuraishi, S., & Okumura, F. S., *Nature, Lond.*, 1961, **189,** 148
[9] Loeffler, J. E., & van Overbeek, J., in 'Régulateurs naturels de la croissance végétale, 1963, No. 123, p. 77 (Paris: C.N.R.S.)
[10] Letham, D. S., in 'Biochemistry and physiology of plant growth substances', in press (Ottawa: Ringe Press)
[11] Letham, D. S., *Life Sci.*, 1963, **2,** 569
[12] Letham, D. S., *Phytochemistry*, 1966, **5,** 269
[13] Miller, C. O., in 'Modern methods of plant analysis', 1963, **6,** 194 (Eds. K. Paech & M. V. Tracey) (Berlin: Springer Verlag)
[14] Kuraishi, S., *Sci. Pap. Coll. Gen. Educ., Tokyo*, 1959, **9,** 67
[15] Hillman, W. S., *Science, N.Y.*, 1957, **126,** 165
[16] Letham, D. S., *Planta*, 1967, **74,** 228
[17] Osborne, D. J., & McCalla, D. R., *Pl. Physiol., Lancaster*, 1961, **36,** 219
[18] Kende, H., *Science, N.Y.*, 1964, **145,** 1066
[19] Kende, H., *Proc. natn. Acad. Sci., U.S.A.*, 1965, **53,** 1302
[20] Gorton, B. S., Skinner, C. G., & Eakin, R. E., *Arch. Biochem. Biophys.*, 1957, **66,** 493
[21] Letham, D. S., *A. Rev. Pl. Physiol.*, 1967, **18,** 349
[22] Bamberger, E., & Mayer, A. M., *Science, N.Y.*, 1960, **131,** 1094
[23] Köhler, K. H., & Conrad, K., *Biol. Rundschau*, 1966, **4,** 36
[24] Conrad, K., & Köhler, K. H., *Wiss. Z. Univ. Rostock, Math.-Nat.*, 1967, **16,** 657
[25] Bigot, C., *C.R. hebd. Seanc. Acad. Sci., Paris*, 1968, **266,** 349
[26] Nitsch, J. P., & Nitsch, C., *Bull. Soc. bot. Fr.*, 1965, **112,** 11
[27] Letham, D. S., *Life Sci.*, 1966, **5,** 551
[28] Koshimizu, K., Matsubara, S., Kusaki, T., & Mitsui, T., *Agric. Biol. Chem.*, 1967, **31,** 795
[29] Helgeson, J. P., & Leonard, N. J., *Proc. natn Acad. Sci., U.S.A.*, 1966, **56,** 60

[30] Hall, R. H., Robbins, M. J., Stasink, L., & Thedford, T., *J. Am. chem. Soc.*, 1966, **88,** 2614
[31] Zachau, H. G., Dütting, D., & Feldmann, H., *Angew. Chem. Int. Edn.*, 1966, **5,** 422
[32] Robins, M. J., Hall, R. H., & Thedford, R., *Biochemistry*, 1967, **6,** 1837
[33] Hall, R. H., Csonka, K., David, H., & McLennan, B., *Science, N.Y.*, 1967, **156,** 69
[34] Klämbt, D., Thies, G., & Skoog, F., *Proc. natn. Acad. Sci. U.S.A.*, 1966, **56,** 52
[35] Nitsch, C., *Ann. Sci. Nat., Bot.*, 12e *Ser.*, 1968, **9,** 1
[36] Wood, H. N., in 'Régulateurs naturels de la croissance végétale', 1963, No. 123, p. 97 (Paris: C.N.R.S.)
[37] Cavé, A., Deyrup, J. A., Goutarel, R., Leonard, N. J., & Monseur, X. G., *Ann. Pharm. fr.*, 1962, **20,** 285
[38] Leonard, N. J., & Deyrup, J. A., *J. Am. chem. Soc.*, 1960, **82,** 6202
[39] Beauchesne, G., & Goutarel, R., *Physiologia Pl.*, 1963, **16,** 630
[40] Rogozinska, J. H., Helgeson, J. P., & Skoog, F., *Physiologia Pl.*, 1964, **17,** 164
[41] Rogozinska, J. H., *Bull. Acad. pol. Sci., Cl. II, Ser. Sci. biol.*, 1967, **15,** 313
[42] Miller, C. O., *Proc. natn. Acad. Sci., U.S.A.*, 1965, **54,** 1052
[43] Letham, D. S., *Life Sci.*, 1966, **5,** 1999
[44] Beauchesne, G., Leboeuf, M., & Goutarel, R., in 'Régulateurs naturels de la croissance végétale', 1963, No. 123, p. 119 (Paris: C.N.R.S.)
[45] Hall, R. H., *Biochemistry*, 1964, **3,** 769
[46] Chibnall, A. C., *New Phytol.*, 1954, **53,** 31
[47] Kulaeva, O. N., *Fiziol. Rast.*, 1962, **9,** 229
[48] Engelbrecht, L., *Flora*, 1964, **154,** 57
[49] Vardjan, M., & Nitsch, J. P., *Bull. Soc. bot. Fr.*, 1961, **108,** 363
[50] Nitsch, J. P., *Ann. N.Y. Acad. Sci.*, 1967, **144,** 279
[51] Nitsch, J. P., & Nitsch, C., *Bull. Soc. bot. Fr.*, 1965, **112,** 1
[52] Itai, C., & Vaadia, Y., *Physiologia Pl.*, 1965, **18,** 941
[53] Seth, A., & Wareing, P. F., *Life Sci.*, 1965, **4,** 2275
[54] Carr, D. J., & Burrows, W. J., *Life Sci.*, 1966, **5,** 2061
[55] Bui Dang Ha, D., & Nitsch, J. P., *C.R. hebd. Séanc. Acad. Sci., Paris,* 1968, in press
[56] Richmond, A. E., & Lang, A., *Science, N.Y.*, 1957, **125,** 650
[57] Mothes, K., Engelbrecht, L., & Kulajerra, O., *Flora*, 1959, **147,** 145
[58] Leopold, A. C., in 'Régulateurs naturels de la croissance végétale', 1963, No. 123, p. 705 (Paris: C.N.R.S.)
[59] Mothes, K., in 'Regulateurs naturels de la croissance végétale', 1963, No. 123, p. 131 (Paris: C.N.R.S.)
[60] Wollgiehn, R., & Parthier, B., *Phytochemistry*, 1964, **3,** 241
[61] Madison, J. T., Everett, G. A., & Kung, H., *Cold Spring Harbor Symp. Quant., Biol.* 1966, **31,** 409
[62] Fox, J. E., *Pl. Physiol., Lancaster*, 1966, **41,** 75
[63] Fox, J. E., & Chen, C., *J. biol. Chem.*, 1967, **242,** 4490
[64] Jablonski, J. R., & Skoog, F., *Physiologia Pl.*, 1954, **7,** 16
[65] Nitsch, J. P., & Bui Dang Ha, D., *C.R. hebd. Séanc. Acad. Sci., Paris,* 1967, **264**(D), 288
[66] Nitsch, J. P., Colloq. C.N.R.S. on Cultures de tissus de plantes, Strasbourg, France, in press
[67] Nitsch, J. P., in 'Biochemistry and physiology of plant growth substances', in press (Ottawa: Ringe Press)
[68] van Overbeek, J., Loeffler, J. E., & Mason, I. R., *Science, N.Y.*, 1967, **156,** 1497
[69] Stern, H., in 'Régulateurs naturels de la croissance végétale, 1963, No. 123, p. 19 (Paris: C.N.R.S.)
[70] Nitsch, J. P., & Bui Dang Ha, D., *C.R. hebd. Séanc. Acad. Sci., Paris,* 1968, in press

[71] Hall, R. H., in 'Biochemistry and physiology of plant growth substances', in press (Ottawa: Ringe Press)
[72] Henderson, T. R., Skinner, C. G., & Eakin, R. E., *Pl. Physiol., Lancaster*, 1962, **37,** 552
[73] Skoog, F., & Miller, C. O., *Symp. Soc. Exp. Biol.*, 1957, **11,** 118
[74] von Maltzahn, K. E., *Nature, Lond.*, 1959, **183,** 60
[75] Szweykowska, A., *J. exp. Bot.*, 1963, **14,** 137
[76] Bopp, N., & Brandes, H., *Planta*, 1964, **62,** 116
[77] Nitsch, J. P., Bui Dang Ha, D., & Nitsch, C., *Bull. Soc. bot. Fr.*, 1966, **113,** 425
[78] Heide, O. M., & Skoog, F., *Physiologia Pl.*, 1967, **20,** 771
[79] Nitsch, J. P., Nitsch, C., Rossini, L. M. E., & Bui Dang Ha, D., *Phytomorphology*, in press

DISCUSSION

Professor R. L. Wain.—I wonder whether Professor Nitsch would like to speculate on the site and mode of action of cytokinins within the plant. Is the substituent at the 6-position of the adenine molecule required to confer favourable physical properties so enabling the adenine moiety to be transported to the site of action where free adenine is then released? If so, would Professor Nitsch like to comment on the report that certain derivatives of urea with no obvious structural relationship to adenine, possess cytokinin activity?

Professor Nitsch.—Apparently, only 6-substituted adenines are active as cytokinins. Reported activities of 1-substituted adenines have been explained by Skoog and co-workers as resulting from a shift of the side-chain from the 1- to the 6-position. It is difficult to say yet that the side-chain is just a handle by which enzymes take hold of an adenine molecule to insert it into a chain as adenine, breaking off the handle later on. The fact that natural cytokinins are found in t-RNA's with their side-chains intact, argues against this hypothesis.

At the present time I have no explanation for the reported activity of substituted ureas.

Professor J. van Overbeek.—If I recall correctly, at the Ottawa conference, last July, Carlos Miller stated that adenine itself has definite cytokinin activity. Therefore the side-chain is not necessary for activity; it may enhance activity.

THIRD SESSION

Chairman: Prof. P. F. WAREING

HORMONAL CONTROL OF FRUIT DEVELOPMENT, SEED DORMANCY AND GERMINATION WITH PARTICULAR REFERENCE TO *ROSA*

By G. A. D. JACKSON*

(*School of Plant Biology, University College of North Wales, Bangor*)

The hormonal requirements for fruit development appear to vary in different species of rose. Thus, parthenocarpy may be induced by the application of auxins to emasculated flower-buds of *R. rugosa*, while *R. arvensis* is indifferent to the treatment. Gibberellin A_3 is effective with these and all other species tested, but although the parthenocarpic hips of *R. arvensis* so produced are similar in size to normal, fertilised hips, this is achieved solely by cell enlargement in the flesh; cell division ceases after emasculation. Auxin- and gibberellin-like activity has been detected in extracts of growing hips, and the possibility is considered that at flowering the receptacle is in one part or another of a 'gibberellin phase' of development, i.e. when gibberellin is either a contributory or dominant factor in the hormonal control of its growth, and the tissues are responsive to gibberellin. In *R. arvensis* a cell division factor appears to be also necessary to produce the normal complement of cells in the hip flesh, while in *R. rugosa* the receptacle may be in a later part of the gibberellin phase where it overlaps fully with a succeeding auxin phase. It is suggested that the interplay of hormones arising from the seed, and coming from the parent, may control fruit development and correlate the growth of the fruit with the events occurring in the seed and in the rest of the plant.

Abscisic acid is present in both flesh and achenes of mature hips and is implicated as one of the factors in seed dormancy in *Rosa*. For example, its removal from the excised embryo permits rapid germination, and its effect may be annulled by gibberellin A_3, or the synthetic cytokinin 6-benzylaminopurine but not by 3-indolylacetic acid. The state of the embryo may therefore be controlled by the balance between abscisic acid on the one hand and gibberellin or cytokinin (or both) on the other. The presence of the β-inhibitor complex in all stages in the growth of hips so far examined, together with the possibility that abscisic acid may induce some parthenocarpic development in one species of rose, suggests that the balance between abscisic acid and the growth promoting hormones may be important throughout the whole life of the hip.

*This paper was presented at the Symposium by Dr. J. B. Blundell, since the author was unfortunately indisposed.

Introduction

In the higher plants, the sexual production of a new individual falls readily into three phases. From the inception of flower primordia to the production of flowers constitutes Phase I. The following Phase II comprises fertilisation and the growth of the fruit containing the embryo, and here the term fruit will generally be used not in the morphological sense with all its accompanying complexities of classification, but physiologically as defined by Nitsch,[1] namely 'A fruit consists of those tissues which support the ovules of a plant, and whose development is dependent upon physiological changes occurring in these ovules'. The final Phase III consists of the transition of the embryo into an actively-growing, individual plant.

This all-too-easy separation into three phases, of what is essentially a continuous process, has tended to restrict any given experimental investigation to only one of the phases. As part of a long-term project in these laboratories, investigations are being carried out on the hormonal control of Phases II and III in the roses, i.e. the development of the fruit followed by germination.

Fruit development

Part played by seeds in controlling fruit growth

It is at the borderline between Phases I and II, when the flower becomes a young fruit, that the first natural check in the continuum may occur (and emphasise the borderline) since, in the absence of fertilisation, the ovule and fruit usually fail to develop although some growth may take place. The classic explanation of this failure of the fruit to develop is implied in Nitsch's definition: 'A fruit consists of those tissues . . . *whose development is dependent upon physiological changes occurring in these ovules.*' That the presence of developing seeds is normally required for fruit growth is well known. Nitsch[2] demonstrated this dependence on a finer scale in the strawberry by removing various patterns of achenes from young fruits. Receptacle growth continued only beneath the remaining achenes and various abnormally shaped fruits were produced. There was also a rough proportionality between the number of developing achenes and the size of the mature fruit. No precise relationship has yet been sought for *Rosa*, but Table I suggests that the thesis probably holds in a very general way.

A relatively infertile bush 'A' of *R. rugosa* which set few seeds (or, more correctly, achenes, see page 130), was crossed with pollen from a fertile bush 'B' which produced an abundance of achenes and larger hips. Both the achene set and hip size were

Table I

Effect of achene number on hip size in R. rugosa

	Bush 'A'	Bush 'B'	Bush 'A' ♀ × Bush 'B' ♂
Average hip diameter, mm	23	29	28
Average number of achenes per hip	15	153	130

increased in the cross, but not proportionately. A similar situation was found when a relatively infertile bush of *R. rugosa* was crossed with pollen from *R. spinosissima*. The number of achenes per hip increased from 14 to 105; this was accompanied by an increase in hip diameter from 22 mm to 30 mm, which represented an increase in hip weight of ~60%. Occasionally, however, only one large achene has been found in medium-sized hips of *R. spinosissima*, and more frequently in hips of two natural hybrids, *R. mollis* × *R. spinosissima* and *R. dumalis* × *R. spinosissima*. This argues against any direct proportionality, and suggests that the effect per seed or achene is disproportionately great when only a few are present.[3]

Use of induced parthenocarpy as a means of identifying hormones produced by the developing seed

The classical view further suggested that the developing ovules promote the growth of the supporting tissues by exporting growth hormones, so that if ovule development fails completely through lack of fertilisation, or for any other cause, the supply of hormones is cut off and the fruit does not grow. Later developments in the view will be discussed on page 152, but whatever the mechanism by which these hormones act it is likely that they are vitally involved in controlling the overall growth of fruit and seed.

The check brought about by lack of fertilisation may be exploited. By emasculating the flower at the bud stage, ovule and seed development is prevented and the natural production of growth regulators is eliminated; one may then seek to replace them by external application and so determine their nature. In other words, to use a sort of natural tissue culture in which, with the rose, the hip receives its supplies of water, metabolites, mineral nutrients and probably some hormones from the parent plant, but is dependent upon external supplies of regulators for successful growth into parthenocarpic fruit. Until about 1957, attempts to produce parthenocarpic fruits revolved around the use of auxins and, outside the Solanaceae and Cucurbitaceae, met with little success. However, on examination of the rose family, it was found that it appeared to be a sort of microcosm of the higher plant world in this respect, giving within its species the same range of response to auxins—from sensitivity to indifference—which had been reported for the higher plants as a whole.

It was perhaps fortunate that these experiments began with the Asian species, *R. rugosa*, which proved to be responsive to auxins.[4] The unopened flower bud was emasculated by cutting off the top at the 'disc', so removing calyx, corolla, stamens and stigmas. The auxin was then applied to the cut surface as a drop of aqueous solution or as a lanolin paste, or injected in aqueous solution into the cavity of the receptacle. Nine different auxins were used at different concentrations and best results were obtained with 1% NAAM (1-naphthylacetamide) in lanolin. Parthenocarpic hips similar in outward appearance and approaching the size of normal, fertilised ones were produced (Fig. 1a). IAA (3-indolylacetic acid) stimulated appreciable fruit development but no hips survived to maturity. When the first parthenocarpic specimens were cut open they were, however, found to contain even more 'pips' than the normal ones (Fig. 1b). In addition to stimulating the growth of the flesh, the auxin had caused development in almost the entire carpel complement of the hip.

Here, reference needs to be made to morphological structure. The rose hip is technically a false fruit in which the lower part of the hollow flesh is probably receptacular in origin, derived by invagination of the floral axis, the hump at the bottom of the cavity being the true apex.[5] The upper part of the flesh in this interpretation is appendicular and derived by fusion of floral members, while the 'pips' are morphologically the one-seeded fruits or achenes, each of which develops from a separate ovary consisting of a single carpel. A longitudinal section of an air-dried, normal, mature achene is shown in Fig. 2. Sectioning a 'pseudo-achene' from a mature, parthenocarpic hip showed that the carpel wall had developed into the thick, woody pericarp in the usual way, while the integuments had produced an apparently normal, two-layered testa. Instead of being filled with the embryo the latter contained, however, only the convoluted and degenerate remains of the nucellus. Nitsch[3] has recorded a number of cases where the application of growth regulators has produced parthenocarpic fruits in which there has been development of the outer parts of the seed. He suggests that the nucellus and seed coats, stimulated to grow by the applied regulator, may in turn continue to produce

(a)

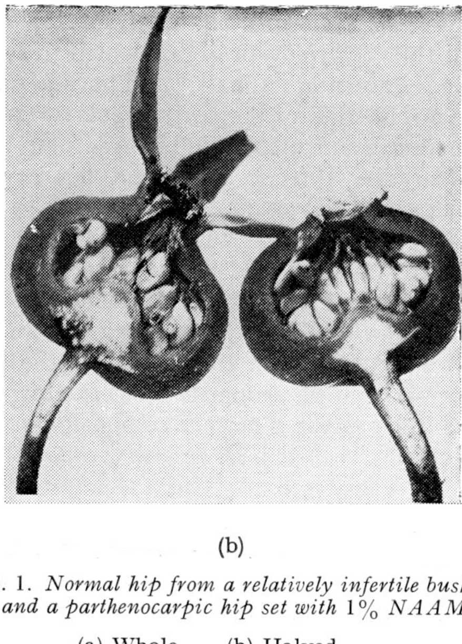

(b)

FIG. 1. *Normal hip from a relatively infertile bush of* R. rugosa, *and a parthenocarpic hip set with* 1% *NAAM in lanolin*

(a) Whole (b) Halved

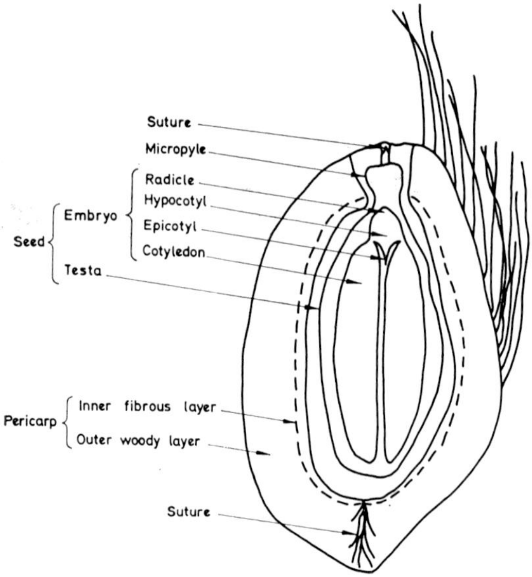

FIG. 2. *Diagram of a median longitudinal section through a mature, air-dried achene of a dormant species such as* R. canina

substances activating fruit growth. This sort of stimulation by one hormone of the growth of an organ, which in turn produces a second hormone is known in animals,[6] and may be the explanation of how a single application of a growth regulator at flowering can carry the parthenocarpic fruit through to maturity.[3]

R. spinosissima, the 'Burnet Rose', did not respond quite so strongly to auxin as did *R. rugosa*, while others such as *R. sherardii* gave a relatively weak response, but the occasional small, mature, parthenocarpic hip was produced in the latter with 0·1% NAAM. At the end of this series *R. arvensis*, the 'Field Rose', was virtually indifferent to auxin in that no ripe fruit was produced, even with concentrations of NAAM down to 0·01%. At that time the gibberellins were thought to be primarily concerned with cell enlargement and, since this process is largely responsible for fruit growth, GA_3 (gibberellic acid) was used. *R. arvensis* responded strongly,[7] as did all the other species tested.

This led subsequently to a re-investigation of parthenocarpic response, using both auxin and gibberellin, and mixtures of the two; the two extreme examples, *R. rugosa* and *R. arvensis*, will be given. Fig. 3 shows the results with a particular bush of *R. rugosa*, which was completely infertile and never produced normal fruit.

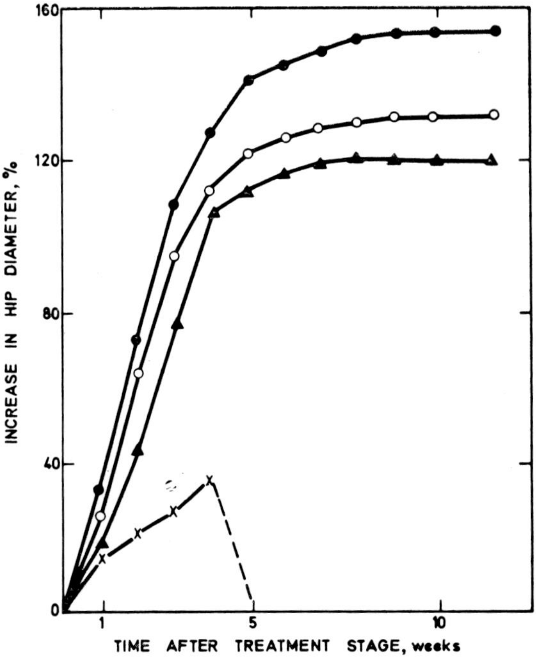

FIG. 3. *Growth of normal and parthenocarpic hips of an infertile bush of* R. rugosa

Curves for the parthenocarpic hips are based only on those individuals which persisted to maturity

× normal hips

▲ emasculated buds treated with 1% NAAM in lanolin

○ emasculated buds treated with 1% GA_3 in lanolin

● emasculated buds treated with 1% NAAM+1% GA_3 in lanolin

Both 1% NAAM and 1% GA_3 in lanolin were successful, but the mixture of the two was even more so and gave an increase in hip diameter at maturity of ~150% compared with ~165% for normal hips of other fertile bushes. Since there is usually some mortality during fruit development, growth curves of this sort should be based only on those fruits which reach maturity. There is no way of knowing in advance which fruits will survive, so a separate series of measurements must be kept for each tagged bud. Failure to do this can produce a point of inflexion (Fig. 4); an apparently double sigmoid curve could even result, owing to a premortem depression in growth rate in those individuals in the sample which would drop in a subsequent wave of abscission. The

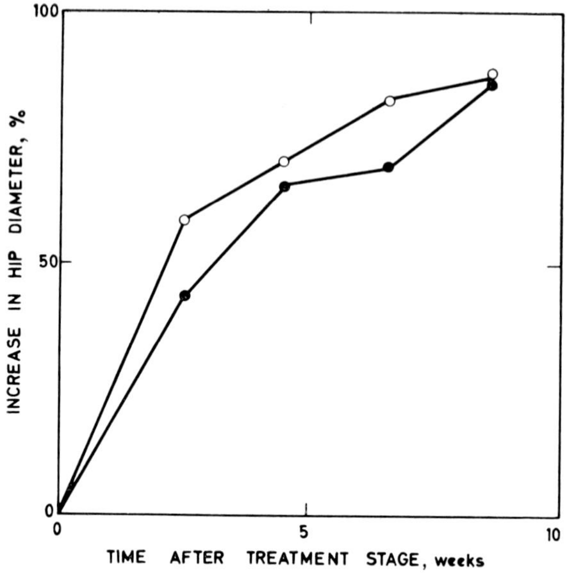

FIG. 4. *Production of a spurious point of inflexion in a growth curve*
Both curves are for the same group of tagged buds of
R. sherardii set parthenocarpically with 0·1 μg of GA_3 per bud
● average diameter of all tagged hips persisting at the time of measurement
○ average diameter of those hips which will persist to the time of maturity

true pattern of fruit growth is being obscured by the premature abscission.

The efficiency of a fruit-setting agent thus depends not only on the size of fruit it produces, but also on the percentage of the treated buds which survive to maturity. Table II shows the superiority of GA_3 over NAAM in this respect, although the figure for the latter treatment may be improved. This is because *R. rugosa* has a much longer flowering period than the indigenous roses, and the time at which the NAAM is used markedly affects the parthenocarpic response; treatment during the optimum period in June can give a 90% survival.

In *R. arvensis* (Fig. 5) the auxin did stimulate more growth than occurred in the emasculated controls which received lanolin alone;

Table II

Percentage survival to the time of maturity of hips from normal flowers of R. rugosa *and* R. arvensis, *and from emasculated buds treated with growth substances in lanolin.* Results are from the same experiments as those on which Figs 3 & 5 are based

Treatment	% of flowers or buds surviving to the time of maturity	
	R. rugosa	*R. arvensis*
Normal flowers	0	89
Buds+lanolin only	—	13: none ripened
Buds+1% NAAM	37	25: none ripened
Buds+1% GA_3	93	92
Buds+mixture containing 1% NAAM +1% GA_3	86	98

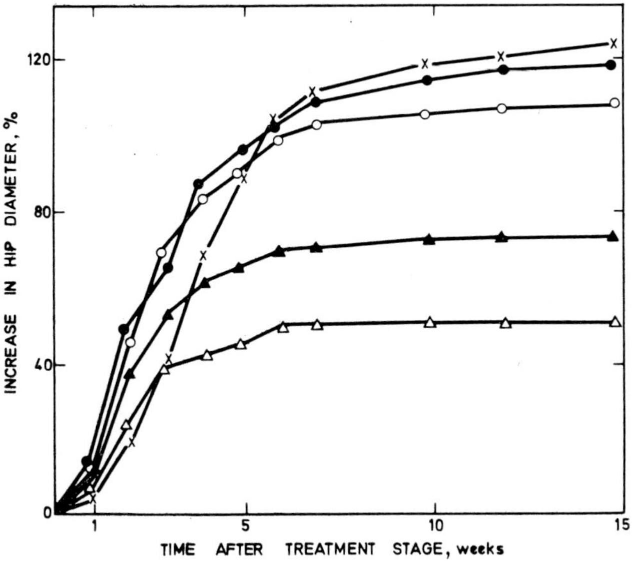

FIG. 5. *Growth of normal and parthenocarpic hips of* R. arvensis
Curves are based only on those individuals which will persist to the time of maturity

× normal hips
△ emasculated buds treated with lanolin only
▲ emasculated buds treated with 1% NAAM in lanolin
○ emasculated buds treated with 1% GA_3 in lanolin
● emasculated buds treated with 1% NAAM+1% GA_3 in lanolin

of the 25% of NAAM-treated buds which persisted to term, none however ripened (Table II). GA_3 was again notably successful in promoting fruit growth and survival; the addition of NAAM gave some extra growth although it is not always so with this species.

The method of 'natural tissue culture' therefore suggests that both auxin and gibberellin may be involved in fruit development in the roses; the genus shows a spectrum of response ranging from *R. rugosa* where the provision of either promotes parthenocarpic flesh and pseudo-achene development, to *R. arvensis* where there is an absolute requirement for gibberellin for the production of mature fruit, and this need cannot be met by auxin. In all the species tested, gibberellin was more effective than auxin in stimulating fruit growth, and particularly effective in promoting survival. The use of both hormones together usually gave a positive interaction in growth of greater or less extent. One of the most striking examples of this was obtained with *R. spinosissima* flowering in the early Spring in the greenhouse (Table III). Under these conditions the performance of the normal flowers was relatively

Table III

Development of normal (self-pollinated) flowers of R. spinosissima *and of emasculated buds treated with growth substances in lanolin*

Treatment	% increase in hip diameter at maturity	% survival to maturity
Normal flowers	175	56
Buds+1% NAAM	164	50
Buds+1% GA_3	198	67
Buds+1% NAAM & 1% GA_3	264	100

poor, since they had to rely on self-pollination and are only partly self-fertile. In the parthenocarpic hips, both the percentage survival and hip diameter were greater with the GA_3/NAAM mixture than with either growth substance alone. Bearing in mind that the volume of a hip varies as the cube of the radius, and considering also the percentage survival, the total yield of fruit from the mixture was truly superior to that provided by either GA_3 or NAAM alone.

The effectiveness of GA_3 and NAAM applied together was also shown by their ability to produce mature hips of *R. rugosa*, when the gynoecium was scooped out at the bud stage and replaced by lanolin paste containing the mixture. Although the percentage survival was reduced by this major, surgical operation and the hips so produced were not as large as those formed by the normal method of application, any contribution of the sort suggested by Nitsch,[3] in the shape of growth activators exported by the developing pseudo-achenes, is not essential for flesh growth in the presence of both growth substances in this species. Such activators may however make some contribution, particularly if only one growth regulator is applied. Either NAAM or GA_3 applied to the cut surface of emasculated buds produces an abundance of pseudo-achenes and forms mature hips. Yet excavated buds treated as above with NAAM alone had all dropped off by week 3, while those supplied similarly with GA_3 alone produced smaller hips than the mixture which did not ripen properly. A similar overall picture emerged when the growth substances in lanolin were applied to the outside of the receptacle of emasculated buds. With the mixture there was virtually no pseudo-achene development, yet the increase in hip diameter and percentage survival were equal to that when the lanolin paste was applied to the cut surface of the 'disc' in the usual way, with the resultant production of pseudo-achenes. But again, although the application in this way of either GA_3 or NAAM alone produced the occasional, mature hip, the percentage survival was very low.

There is thus a strong case for including gibberellin and auxin (and cytokinin; page 152), both as mixtures and in sequence, in trials seeking to promote optimum fruit production where this is held back by infertility or an environment unfavourable for pollination. For example, in the sweet cherry 'Bing', a mixture of gibberellin and auxin produced parthenocarpic development even although neither alone was effective.[8]

Use of gibberellic acid in the setting of difficult crosses in Rosa

In a practical application of the above results, some assistance was given to Allen[9] in the setting of difficult, horticultural crosses where pollen-tube growth may be slow and abscission occur before fertilisation is achieved. GA_3 was most effective in preventing abscission and in the following crosses, three drops of a 250 ppm solution of GA_3 in 25% isopropanol were applied to the base of the exposed part of the stigmas 10–14 days after pollination. If the growth substance is applied too soon after pollination, normal achene development is suppressed and a parthenocarpic fruit is produced. In the first cross, 'Goldilocks' ♀ × (*R. ecae* × 'Canary Bird') ♂, the female parent was tetraploid and the male was diploid. With this wide cross, the gibberellin treatment of a given number of flowers gave a nine-fold increase in the number of viable embryos as compared with the same number of control flowers. The second cross was a less difficult one between two tetraploid varieties, 'Elsinore'♀ × *R. gallica versicolor* ♂, but even here the treatment yielded a four-fold increase in viable embryos.

Pollen incompatibility is, however, a complex phenomenon and it is possible that the gibberellin had other effects as well as preventing abscission. For example, Lloyd[10] believes that activation of pollen growth, following uptake and utilisation of a stigmatic growth substance, could explain many of the features of pollen tube growth in his incompatibility studies in *Leavenworthia*. The nature of such promoters appears to be very varied; Bose[11] found that certain concentrations of GA_3 stimulated the tube growth of isolated pollen grains of *Pisum sativum* but, in a comprehensive survey, Chandler[12] obtained deleterious effects in many species, including the Floribunda rose 'Betty Pryor'. The latter result may explain why the too-early application of GA_3 suppresses normal achene development in the roses, and adds weight to the original suggestion that it probably acts by preventing premature abscission. The suppressive effect of a too-early application of GA_3 also argues against the idea that the incompatibility is due to inhibitors in the stylar tissue,[13] which may be annulled by the applied gibberellin.

Auxins have also been used to overcome incompatibility in other genera; for example, in *Rosa*'s neighbouring genera, *Pyrus* and *Malus*, pear-apple intergeneric hybrids were obtained by the use of 2-naphthoxyacetic acid.[14]

Significance of growth substances extracted from developing hips

The involvement of auxin and gibberellin in development was again suggested by preliminary analyses of extracts of normally developing hips. With the wheat coleoptile, straight-growth assay, a number of zones of stimulation were obtained from chromatograms of the neutral and acidic fractions (separated on an activated alumina column) of extracts made in diethyl ether from hips of *R. rugosa*, *R. spinosissima*, *R. dumalis* and *R. arvensis*. An active zone common to them all occurred in the acid fraction at an R_f

corresponding to that of IAA in two different solvent systems.[7,15,16] This has little value as an identification,[17] but it is of somewhat more interest that Fig. 6 shows a rough correlation in *R. arvensis* between the rise and fall in the auxin content per hip, and the onset and decline of the grand period of hip growth. A similar pattern was obtained for *R. rugosa*, although the auxin peak was broader. On a concentration basis, the auxin built up more quickly and in *R. arvensis* a trough appeared at eighteen days after flowering, prior to which there had been a slight falling off in the rate of growth. Since the latter necessarily referred to the entire sample taken for analysis and not just to hips which would persist to maturity (page 133), both the falling off and the trough may have been related to premature abscission, which began about that time.

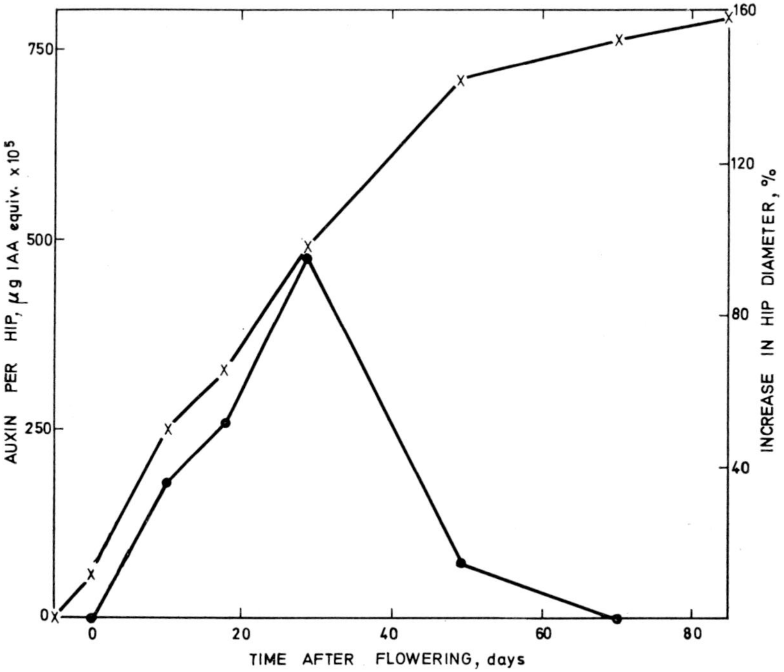

FIG. 6. R. arvensis: *changes in average diameter and in an extractable auxin in samples of normal hips taken at different stages of growth*

× percentage increase in hip diameter

● content per hip of an auxin with an R_f similar to that of IAA as determined by the wheat coleoptile, straight-growth test

The obvious deduction from such apparent correlations must, however, be accepted with reserve. For example, Wright[18] has shown what appears to be a direct relationship between the rate of endosperm and embryo growth, two peaks in the concentration of a particular auxin, and two surges of growth in the blackcurrant: in peach, however, the peaks in concentration of three auxins were associated with the periods of least growth.[19] In addition to the uncertainties attending auxin extraction and bioassay, this sort of interpretation of the size of the auxin pool in a tissue takes little account of the dynamic balance involved. Growth substances may accumulate in an organ (as they and other substances do in developing seeds) because they are not being consumed in growth and translocated away with sufficient rapidity. Conversely, the amount may be low when consumption or translocation are more active; Wain found no IAA in rapidly growing apices of runner bean, hop and grape vine.[20]

A curious result was obtained in a small experiment when the natural growth promoter running in the IAA position was applied to emasculated buds of *R. arvensis* (which cannot be set with synthetic IAA or other auxins). The promoter was separated chromatographically from the developing achenes of five hundred hips harvested at four to five weeks after flowering, and injected in aqueous solution into the cavity of ten emasculated buds of the same species. Three of the buds were subsequently removed by extraneous biotic factors, but the remaining seven grew to maturity and increased in diameter by 98% as compared with 116% for the normal, fertilised controls. A certain amount of gibberellin comes through when wet ether is used as the extractant, and may even affect coleoptile assays.[16] Although no gibberellin activity was detected by the wheat first-leaf test at the R_f of the promoter, it is always possible that a natural gibberellin was missed by this test and that the gibberellin shown to be essential in earlier experiments was in reality being applied.

The wheat first-leaf test was followed by the Meteor pea assay on chromatographed extracts of immature, normal hips.[16] Gibberellin activity was found in all five species tested but the only common zone of activity came, however, at an R_f slightly less than that of GA_8 in n-butanol/ammonia/water: 10/1/1. Bearing in mind the relative insensitivity of the Meteor assay to GA_8, it was initially supposed that the gibberellin was not likely to be this one, particularly after an attempt was made to set hips of *R. spinosissima* and *R. sherardii* parthenocarpically with six gibberellins.[21] None of the buds treated with GA_8 survived, but the other gibberellins all produced mature fruit; the order of effectiveness in stimulating growth in diameter after a twelve-week period was:

R. spinosissima $A_1 \geqslant A_4 \simeq A_9 > A_3 \geqslant A_5 \gg A_8$

R. sherardii $A_1 \simeq A_3 \geqslant A_4 > A_5 > A_9 \gg A_8$.

The results are in general agreement with the suggestion made by Wittwer & Bukovac[22] that, with the exception of GA_3, differences among these gibberellins in the promotion of fruit growth are quantitative rather than qualitative. The properties of the GA_8-glucoside isolated by Sembdner *et al.*,[23] together with its suggested role in the accumulation of de-activated gibberellin during fruit ripening, make it possible that this was the compound present in the chromatographed extracts.

Measurements of cell number and cell volume in hip flesh, and their significance in relation to hormonal control

Later work[24] suggested indirectly that the hip flesh may require the presence of more than auxin and gibberellin if it is to develop normally. Changes in cell size and cell number were followed during growth in the flesh of normal buds of *R. arvensis*, and of emasculated buds treated with lanolin containing 1% GA_3, 1% NAAM or 1% of both. Fig. 7 records changes in total cell number in the

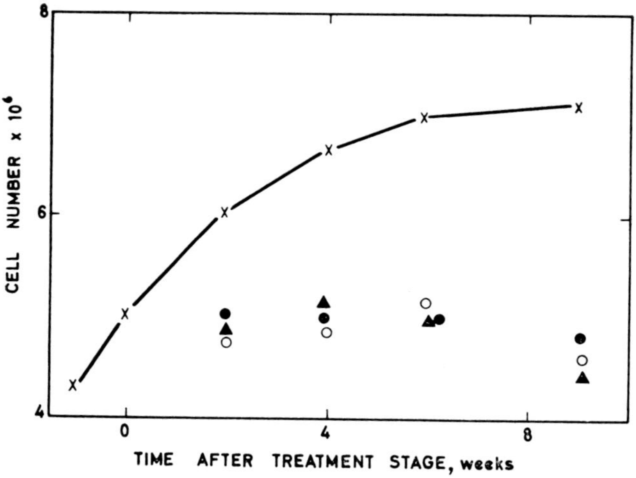

FIG. 7. R. arvensis: *changes in the total number of cells in the flesh of normal and parthenocarpic hips during growth*

× normal hips

▲ emasculated buds treated with 1% NAAM in lanolin

○ emasculated buds treated with 1% GA_3 in lanolin

● emasculated buds treated with 1% NAAM+1% GA_3 in lanolin

flesh during development, and shows that, in the normal flesh, cell division went on for some weeks after flowering, the number of cells rising from about five million to seven million. In the parthenocarpic hips however virtually no cell division occurred in the flesh, even although those set with GA_3 or the mixture would at maturity approach the diameter of normal hips. The explanation of this anomaly in the case of the GA_3-treated hips came directly from the results on average cell volumes (Fig. 8). In the normal flesh, cell volume did not begin to increase appreciably until after week 2, when cell division was slowing down. In the parthenocarpic hips set with GA_3 however the cells were translated immediately into their maximum rate of enlargement, similar to that shown by cells of the normal flesh only after week 4. This rate of enlargement continued almost unabated during the nine weeks of measurement and so these cells ended up much larger than the normal ones; what the parthenocarpic flesh had lost in cell-division it had almost regained in vacuolation.

The cells of the flesh in the hips set with the hormone mixture also enlarged rapidly at first, but then the rate diminished and in the end the cells were little larger than normal. Yet these par-

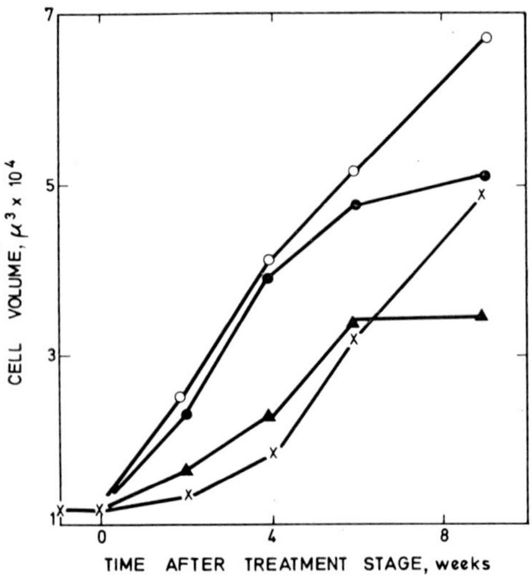

FIG. 8. R. arvensis: *changes in average cell volume in the flesh of normal and parthenocarpic hips during growth*

× normal hips
▲ emasculated buds treated with 1% NAAM in lanolin
○ emasculated buds treated with 1% GA_3 in lanolin
● emasculated buds treated with 1% NAAM+1% GA_3 in lanolin

thenocarpic hips were almost as large in diameter as normal hips (*cf.* Fig. 5), in which the flesh contained two million more cells. The explanation here lay in the inadequacy of such a single, linear measurement as a criterion of fruit size, which really concerns fruit volume. The *length* of the hips set with the mixture was less than that of the normal ones; the hip volume and flesh volume were also less, as might have been expected from the cellular picture.

One interpretation of the findings on cell numbers would be that a cell division factor such as a cytokinin was lacking in the parthenocarpic hips, so that whatever cytokinin came to the flesh in the sap from the parent bush,[25] it could not replace some factor normally provided by pollination or pollen tube growth or produced in the developing seeds. Alternatively, the effect of the cell division factor may have been eliminated by the exogenous growth substances. This sort of antagonism, for example between endogenous gibberellin and applied auxin, has been suggested as the mode of action of the latter in fruit-thinning.[6] The smaller cell and flesh volume ultimately produced by the GA_3/NAAM mixture, as compared with that promoted by GA_3 alone, may have been due to a similar effect. The idea may also explain why lower concentrations of auxin alone were more effective in setting parthenocarpic fruit in *R. sherardii* (page 132), in which the cells of the flesh may be further on in a 'gibberellin phase' of development at flowering than are those of *R. arvensis* (page 151).

Seed dormancy and germination

Natural inhibitors as one of the causes of dormancy, and its relief by growth promoters

As the fruit approaches maturity the second check in the growth of the new individual may begin, emphasising the borderline between Phases II and III. In the roses the embryo becomes dormant and the depth varies with the species, ranging from *R. polyantha* where the achenes will germinate in a matter of weeks, to deeply dormant species such as *R. canina* where the achenes will lie under natural conditions for eighteen months before giving a germination of about 30%.

Fig. 2 showed diagrammatically a median longitudinal section through an air-dried achene of a dormant species such as *R. canina*. The two cotyledons of the apparently mature embryo are stored with oily food reserves and surrounded by a thin, papery testa to give the seed proper. This is in turn enclosed in the thick, hard pericarp or fruit-wall, to give the morphological fruit or achene.

If the embryo is anatomically mature, the question arises as to whether dormancy is imposed on it by the surrounding testa and pericarp or whether the embryo itself is dormant. Naked embryos excised from achenes of all the dormant species so far tested germinated to a greater or lesser extent at 26°, with the exception of *R. sherardii,* which was still dormant after 15 days. Leaching the dormant embryos with water for 5 days, however, produced 100% germination during this period.[26] This was the first hint that the dormancy of the embryo might be due to some water-soluble inhibitor, and was imposed on the embryo by the surrounding tissues.

Another question was whether the pericarp and testa brought about dormancy by acting as a strait-jacket on the embryo, or by denying it access to water or air. Since Tincker[27] found no germination even after cracking the pericarp and soaking in water neither of these ideas seemed likely, and it was confirmed that water could pass through to the embryo in two species. This suggests a return to the notion of diffusible germination inhibitors in the outer parts of the achene, and Fig. 9 shows that in *R. arvensis* the

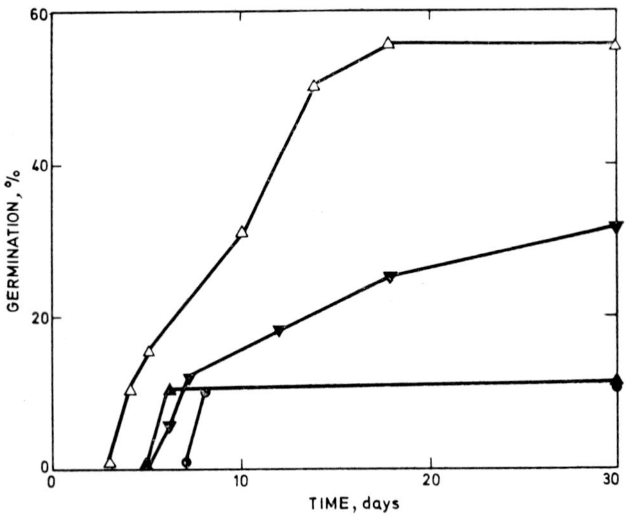

FIG. 9. *Effect of pericarp and testa on germination in* R. arvensis
△ naked embryos
▼ naked embryos mixed with pericarps and testas on the germination pad, but not touching them
▲ naked embryos covered with half-pericarps
● entire seeds (embryo + testa)

mere presence of pericarp and testa near the naked embryos on the germination pad reduced germination. Covering each embryo with a half-pericarp reduced germination still further, in fact to approximately that shown by entire seeds. The inhibitory effect of the testa in the entire seed could be partially relieved by soaking the latter in water for 16 h even in the deeply dormant species, *R. canina*. Soaking the seed of *R. canina* for a similar period in the synthetic cytokinin, BAP(6-benzylaminopurine), at a concentration of 10 μg/ml more than offset the effect of the testa, in so far as these seeds actually germinated more rapidly and to a higher percentage than did untreated, naked embryos. Essentially similar results were obtained with GA_3 at a concentration of 150 μg/ml, but IAA was without effect.

The presence of inhibitors was confirmed by germinating naked embryos on the eluates from chromatogram-segments of an extract of mature achenes (Fig. 10). Germination was completely inhibited around the centre of the chromatogram in the β-inhibitor position, but could be almost entirely restored by BAP, as had been found in the previous *in vivo* experiments with seeds.

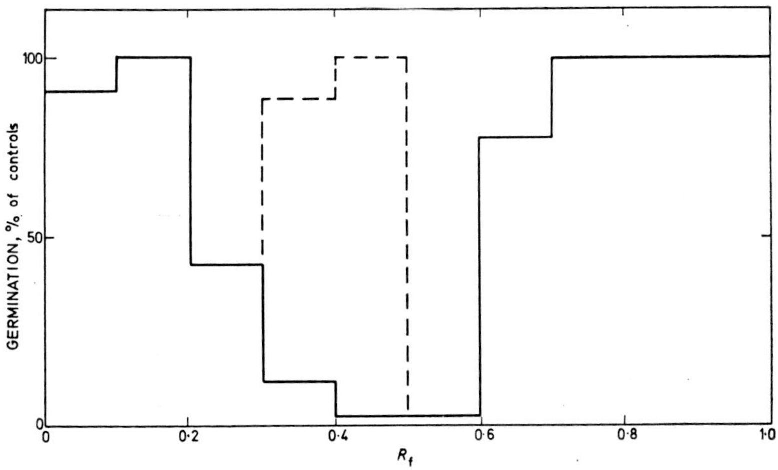

FIG. 10. *Germination after six days of naked embryos in eluates from a chromatogram of the acidic fraction of the ether-soluble substances present in an ethanolic diffusate of dormant achenes of* R. arvensis
Chromatogram run in butanol/ammonia/water: 10/1/1 (Jackson & Blundell[28])
——— segments eluted with water
– – – – segments R_f 0·3–0·5 from a duplicate chromatogram eluted with BAP (10 μg/ml)

Thus the idea developed that it is possibly the balance between germination inhibitors and the active gibberellins or cytokinin (or both) in the embryo which determines its state, i.e. whether it goes into dormancy or germination.

Simulation by abscisic acid of the behaviour of the natural inhibitor

The possibility that the germination inhibitor might be dormin or abscisic acid (ABA) was then tested.[29] For example, naked embryos of *R. rugosa* were imbibed in water for 17 h and then germinated on filter-pads soaked either with water or with ABA solution. Table IV shows that as the concentration of ABA is increased, germination is delayed. When some of the embryos which had *not* germinated on the 1·0 μg/ml solution of ABA by day 5 were removed and rinsed in water, they germinated completely on a pad moistened with water in the next couple of days; thus ABA behaved like the natural inhibitor in not being toxic and being able to be leached out of the embryo. This was even more evident with 50 μg/ml ABA, with which embryos were still dormant after a month, but germinated within 4 days after rinsing with water.

Table IV

Effect of ABA on the germination of naked embryos of R. rugosa

Germinated on ABA: μg/ml	% germination after days					
	3	4	5	6	7	9
0	50	100				
0·01	30	100				
0·1	10	90	100			
1·0	0	0	10	10	70	90
			↓	80	100	

At the point indicated by the arrow, some of the embryos which had not germinated in 1·0 μg/ml of ABA were rinsed in water and placed on a filter pad moistened with water.

Another question was whether BAP and GA_3 could annul the effect of ABA as they did with the natural inhibitor. Embryos were imbibed either by soaking in water for 17 h as a control, or in BAP (10 μg/ml) or in GA_3 (150 μg/ml), and germinated on water or ABA (10 μg/ml). As shown in Table V the water controls, germinated on water, had all germinated by day 4, while those on ABA had only germinated 10% by day 17. Pretreating the

Table V

Effect of pretreatment with GA_3 or BAP on the response of naked embryos of R. rugosa *to ABA*

Pretreatment—embryos imbibed for 17 h in:	Germinated on ABA: μg/ml	% germination after days 4	17
Water	0	100	
Water	10	0	10
GA_3: 150 μg/ml	10	0	100
BAP: 10 μg/ml	10	10	100

embryos with either BAP or GA_3 produced complete germination by day 17 and so ABA was again simulating the behaviour of the natural inhibitor.

A crucial test of the notion that ABA might be acting as an inhibitor *in vivo* was, however, whether it was present in the mature achene. This had first been reported for *R. arvensis* by Cornforth *et al.*[30] In collaboration with Milborrow this work was subsequently extended to other species, and Table VI is a composite record of some of the results.[29,31] If the mature achenes of the three species are first considered, the ABA concentration falls in the series *R. canina*, *R. arvensis* and *R. rugosa*; the depth of dormancy in the embryo follows the same pattern. Thus the germination of naked embryos from the *R. canina* achenes was 35% after 14 days, an average figure for *R. arvensis* would be 65% (there was no opportunity to test the actual sample) while *R. rugosa* germinated completely in four days. Secondly, in a

Table VI

Concentration of ABA in the achenes and flesh of rose hips*

Species	Material	ABA concentration, mg/kg fresh weight
R. canina	Mature achenes	0·53
	After cold stratification	0·05
	After warm stratification	0·16
R. arvensis	Mature achenes	0·16
	Mature flesh	4·1
R. rugosa	Mature achenes	0·058

*Determined by optical rotatory dispersion[31]

sample of the *R. canina* achenes which had been stratified at 2° until they were beginning to germinate, the ABA concentration had fallen to about one-tenth of the original value; achenes which had been stratified for an equal period at 21° remained dormant and contained three times as much ABA as the cold-stratified achenes.

Jacobs[32] erected six criteria to decide whether a naturally occurring substance actually controls a particular process *in vivo*; it seems likely that ABA satisfies these requirements so far as dormancy in *Rosa* is concerned.[31] It would, however, be a mistake to believe that the concentration of ABA in the mature achene is the only factor involved in dormancy. For example, the thickness, texture and composition of the pericarp will affect the rate at which it decays and the rate at which ABA is dissipated in the soil water; it is not even known whether there is just an initial reservoir to be leached from the pericarp or whether further ABA is produced during the natural breakdown of the pericarp. Equally important is the possibility that growth promoters are involved in the balance which determines the state of the embryo (page 146). Work at present in progress on the more detailed changes in ABA and in gibberellins during stratification should further clarify the picture.

Presence of abscisic acid in hip flesh

Table VI also shows that the flesh of mature hips of *R. arvensis* contains a much higher concentration of ABA than is present in the achenes.[31] Although subsequent assays for the flesh of other species have not given such high values, it does invite speculation that perhaps exports of ABA from the ageing flesh of this terminal organ contribute to the onset of dormancy in the embryo. Certainly achenes are slower to germinate if they are sown in the whole hip. Conversely, achenes taken from unripe hips are less dormant;[33] the onset of pigmentation and intensive carotenoid production in the ripening flesh appear to be roughly correlated with the onset of dormancy in the embryo. The possibility of a relationship between pigmentation and the accumulation of ABA has been reported in the strawberry,[34] where the ABA concentration in red, ripe fruit was two-and-a-half times that in white, unripe berries collected at the same time. ABA might be produced in ripening fruits by the transformation of carotenoids[35] or synthesised independently from a common precursor; Rudnicki & Antoszewski (personal communication) favour the latter pathway.

Abscisic acid and fruit development

While investigating the effect of ABA on hip development, it was found that quantities ranging about 1·0 μg applied in aqueous solution to each emasculated bud of *R. sherardii* produced some fruit set and parthenocarpic development.[36] Fig. 11 shows that the growth in hip diameter stimulated by ABA was considerably less than that shown by the normal hips or by those set with 1·0 μg GA_3 per bud. As compared with the number of pseudoachenes per hip produced by GA_3, the set of pseudoachenes with ABA ranged from 0–100% in different hips. In fact the parthenocarpic response to ABA was more like that to auxin, to which *R. sherardii* reacts weakly. By comparison with the emasculated

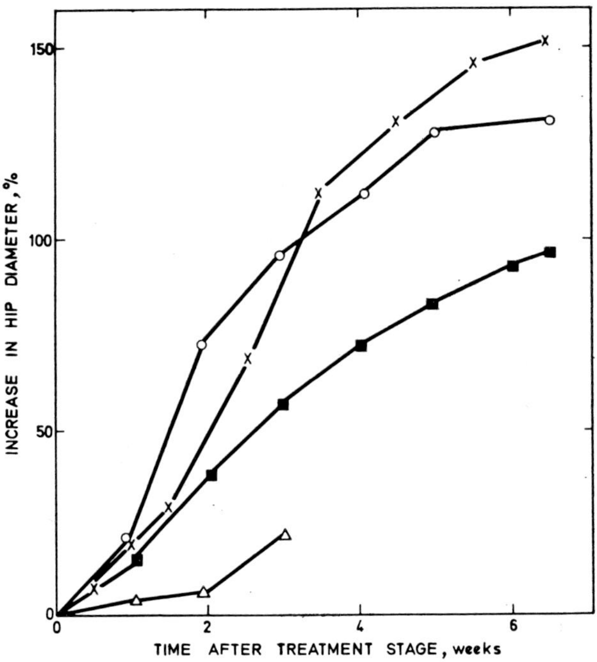

FIG. 11. *Growth of normal and parthenocarpic hips of* R. sherardii (*Jackson & Blundell*[36])

× normal hips

△ emasculated buds treated with lanolin only; the curve here refers to all the tagged hips persisting at the time of measurement, they had all dropped off by week 4

○ emasculated buds treated with GA_3

■ emasculated buds treated with ABA

controls receiving only lanolin the effect, however, appears to be real. No parthenocarpic set has so far been obtained with other species of rose, and keen interest was therefore evoked by the report of Aspinall *et al.* of a stimulation by ABA in the presence of gibberellin of the extension growth of cucumber hypocotyl.[37] This reinforced the suggestion that ABA may have a place with the other plant hormones in the complex balance by which growth in general is controlled.[36]

The β-inhibitor complex is widely distributed in plants—it is found for example in hip tissues of all ages—and evidence is growing that the component in it with functional significance is ABA.[31] It is just possible that at certain times ABA may offset a supraoptimal concentration of a growth promoter, thus appearing to have opposite effects in different systems as does auxin. That an inhibitor can in some situations provoke a growth response is seen with maleic hydrazide, which has been used to set parthenocarpic fruit in the strawberry.[38]

Summary and conclusions

If it is assumed that the hormonal pattern regulating the development of the receptacle and hip flesh is basically similar throughout the genus *Rosa*, the composite results from the different species obtained by the 'natural tissue culture' method of parthenocarpic development suggest the involvement of a cell division factor, gibberellin, auxin and possibly abscisic acid. From work on other fruits it seems likely that, in normal development after flowering, the endosperm and embryo are the most important foci of growth-promoter production in the hip, but other parts of the seed may also be concerned (page 130). Thus it is common enough in certain infertile rose hybrids to find good-sized hips containing one large achene which of course lacks an embryo (page 129). Even in hybrids where some of the hips contain no mature achenes,[39] there has usually been a certain amount of early carpel development before abortion. In an early experiment Harrison[40] removed stigmas and stamens from buds on bushes of *R. canina* var. *lutetiana* and *R. dumetorum*. He records of the hips so produced that at harvest 'not a single fruit contained anything save a few useless chaffy scales'; the scales may again however have been slightly developed carpels. Nitsch[3] in fact suggests that some ovule development is crucial in the growth of naturally parthenocarpic fruits.

For the reasons given on page 140 (which are by no means exhaustive[3]), it does not seem likely that the rise and fall in concentration of the sequence of growth hormones extracted from fruits during development has, necessarily, any direct quantitative relationship with the rate of growth. As evidence for the relative order in which they become pre-eminent in controlling the various phases of fruit development, one may be on somewhat surer ground with the sequence cytokinin, gibberellin and auxin.[3] Certainly cytokinins are widely regarded as regulators of cell division but even here, Wright[41] has found that in the earliest stages of growth excised wheat coleoptiles were most sensitive to applied GA_3, and only later did kinetin produce the greater stimulation of growth. In the case of the flesh of the receptacle it must be remembered, however, that it has already passed through Phase I of its development, and that cell division is mostly (or completely, as in *Lycopersicum esculentum*[42]) over by flowering. Whatever the order in which the different hormones may become predominant in regulating fruit development, there is interplay between them and a considerable (or even total) overlap in their involvement in growth. In general, all three hormones may affect the phases of cell division, enlargement and differentiation; for example the interaction between cytokinin and auxin appears to be highly significant in morphogenesis.

Considering however the sequence cytokinin, gibberellin and auxin, since all species of rose may be set parthenocarpically by GA_3 it seems possible that, at flowering, the receptacle is in one part or another of the gibberellin phase, i.e. when gibberellin is either a contributory or dominant factor in the hormonal control of its growth, and the tissues are able to respond to it. Auxin is at the end of the postulated sequence and this could explain its inability to set parthenocarpic fruit in *R. arvensis*. Here the flesh is still in the process of cell division at flowering (page 141), i.e. may be in the earlier part of the gibberellin phase and still controlled also by cytokinin. In fact when auxin was added to the GA_3 it reduced the final cell and flesh volume produced (page 142).

Preliminary cell counts suggest that in *R. sherardii* the cells of the flesh are at a slightly later stage of development at flowering, and this may be the reason for its weak response to auxin (page 132). The relatively strong response to auxin shown by *R. rugosa* invites further work to determine whether cell division may be almost over in this species by flowering, with the flesh moving towards the end of the gibberellin phase where it overlaps fully with the succeeding auxin one.

The speculation that one hormone, for example gibberellin, may stimulate the development of tissues (such as those of the pseudoachenes in parthenocarpic fruit, pages 130 and 137) which in turn produce another hormone, for example auxin, is still open to proof. Cases have been reported for a variety of organs and tissues in which the application of gibberellin has led to increased levels of either extractable or diffusible auxin over periods as short as 4 h.[43,44] Just as a gibberellin may activate a specific gene pattern which leads to the synthesis of α-amylase in germinating barley,[45] so the cell may be programmed for the production of auxin by this gibberellin or by another member of the ever-increasing family.

Undoubtedly, the application of GA_3 to emasculated buds of *Rosa* always leads to the production of parthenocarpic hips and presumably to the satisfaction, at least in part, of any need for auxin. But even if gibberellin does turn out to be responsible for the promotion of auxin synthesis this cannot be its only function, since in *R. arvensis* auxin is unable to replace the requirement for gibberellin. Its partial ability to do so in *R. rugosa* (Fig. 3 and Table II), where the flesh at flowering may be in a later part of the gibberellin phase, might be interpreted as meaning that the other gibberellin-controlled functions have by then mostly been discharged; the flesh has passed the critical point up to which gibberellin is essential, and after which it is mainly concerned with promoting auxin synthesis or transport (see below). However, the case of the sweet cherry has already been mentioned (page 137) where the production of parthenocarpic fruit requires the addition of auxin to the applied GA_3-spray, and this hypothesis may well fail in the light of later work.

Attention must now be turned to the likelihood that in addition to the hormones produced by the developing seed, a contribution is also received by the fruit from the vegetative parts of the plant since xylem sap contains cytokinin and other hormones.[25] In 1962 van Overbeek issued a relevant reminder that fruit development must be viewed against the background of the activities of the whole plant. And again, that in a preoccupation with hormonal control, it is necessary not to lose sight of the part played by water relations and solute accumulation in the enlarging cells of the growing flesh.[6] There is no doubt that when a fruit is set it becomes a very active centre for the mobilisation of metabolites.[3] Arising from work on parthenocarpic development in fig, Crane[46] assembled evidence which led to the proposal that 'fruit growth is not controlled by hormones emanating from the seeds but, on the contrary, by their capacity to attract metabolites from other regions of the plant with the fruit tissues surrounding the seeds tapping the metabolite supply and acting as storage organs.'

Since parthenocarpic figs could be set with either a synthetic cytokinin, auxin or GA_3, the inference was that the application of any one of them could lead to the supply of either or both of the others, should they be required, from some part of the plant other than the seeds. The latter situation obviously does not obtain in *R. arvensis*, in which auxin is unable to satisfy the demand for gibberellin and produce mature parthenocarpic fruit. Jacobs & Case[47] have also suggested that gibberellin may function as a mobiliser of auxin rather than by directly promoting its synthesis.

There is as yet no reason to suppose however that all of the growth substances synthesised so actively in the fertilised ovules remain entirely 'bottled up' in the developing seed or fructigenic apex. Indeed, it is possible that hormones move out of the seed to establish the gradient which controls the mobilisation of metabolites from the vegetative parts of the plant. Tissue culture experiments suggest that fruit flesh is not generally autotrophic with respect to hormones,[3] and it is probable that the interplay of these substances coming both from seeds and from the parent controls fruit development, and correlates its growth with the events occurring in the seeds and in the rest of the plant. This meeting of the two streams of hormones in the flesh may also be the reason for its ability to compete successfully with the seeds for a share in any hormone-directed mobilisation of metabolites. In addition, the interplay may explain in part the frequent lack of correlation between the rate of fruit growth and the level of growth substances extracted from the seeds or from the whole fruit (page 140).

The presence of abscisic acid seems to be firmly established as one of the factors in seed dormancy in *Rosa*. Its removal from the excised embryo permits rapid germination, and its effect may be annulled by GA_3 or the synthetic cytokinin BAP but not by IAA. The state of the embryo may therefore be controlled by the balance between ABA on the one hand and gibberellin or cytokinin (or both) on the other. The presence of the β-inhibitor complex in all stages of hips so far examined, together with the unexpected ability of ABA to induce a certain amount of parthenocarpic development in one species, suggests that the balance between ABA and the growth promoting hormones may be important throughout the whole life of the hip. Milborrow has reviewed the hypothesis that this sort of interaction between the inhibitor, ABA, and the growth promoters may be the most effective means of providing a fine control of growth;[31] it is not perhaps surprising that evidence is accumulating to show that a common focus at which they all act is gene replication and action leading to protein synthesis, the hallmark of the living organism.[48–54]

Acknowledgements

The author wishes to thank Dr. M. V. Prosser and Dr. J. B. Blundell by whom most of the experimental work was carried out, and Dr. B. V. Milborrow for the abscisic acid assays. He also wishes to acknowledge generous gifts of abscisic acid from the 'Shell' Research Laboratories and of Atlé wheat from Dr. F. Fajersson, and of grants in support of the work over the years from the Royal Society, the Agricultural Research Council, Imperial Chemical Industries, the Association of British Rose Producers and Rose Distribution Limited. The author also wishes to thank Dr. J. B. Blundell for the presentation of this paper at the Symposium.

References

[1] Nitsch, J. P., *Q. Rev. Biol.*, 1952, **27,** 33
[2] Nitsch, J. P., *Am. J. Bot.*, 1950, **37,** 211
[3] Nitsch, J. P., in Ruhland, W., 'Encyclopedia of plant physiology', 1965, **15,** Part 1, p. 1537 (Berlin: Springer-Verlag).
[4] Jackson, G. A. D., & Prosser, M. V., *Naturwissenschaften*, 1959, **46,** 407
[5] Jackson, G., *Am. J. Bot.*, 1934, **21,** 453
[6] van Overbeek, J., Proceedings Plant Science Symposium, Campbell Soup Company, Camden, New Jersey, U.S.A., 1962, p. 37
[7] Prosser, M. V., & Jackson, G. A. D., *Nature, Lond.*, 1959, **184,** 108
[8] Rebeiz, C. A., & Crane, J. C., *Proc. Am. Soc. hort. Sci.*, 1961, **78,** 69
[9] Allen, E. F., *Natn. Rose Soc. Rose A.*, 1967, p. 123
[10] Lloyd, D. G., *New Phytol.*, 1968, **67,** 179
[11] Bose, N., *Nature, Lond.*, 1959, **184,** 1577
[12] Chandler, C., *Contr. Boyce Thomson Inst. Pl. Res.*, 1957, **19,** 215
[13] Iwanami, Y., *J. Yokohama Munic. Univ.*, 1959, **116,** 1
[14] Crane, M. B., & Marks, E., *Nature, Lond.*, 1952, **170,** 1016
[15] Prosser, M. V., Thesis, Univ. Wales, 1959
[16] Jackson, G. A. D., & Prosser, M. V., *Naturwissenschaften*, 1961, **48,** 649
[17] Steward, F. C., & Shantz, E. M., *A. Rev. Pl. Physiol.*, 1959, **10,** 379
[18] Wright, S. T. C., *J. hort. Sci.*, 1956, **31,** 196
[19] Stahly, E. A., & Thompson, A. H., *Bull. A-104 Md agric. Exp. Stn.*, 1959
[20] Wain, R. L., *S.C.I. Monogr. 31*, 1968, p. 3
[21] Jackson, G. A. D., & Blundell, J. B., *Nature, Lond.*, 1964, **202,** 1027
[22] Wittwer, S. H., & Bukovac, M. J., *Am. J. Bot.*, 1962, **49,** 524
[23] Sembdner, G., Weiland, J., Aurich, O., & Schreiber, K., *S.C.I. Monogr. 31*, 1968, p. 00
[24] Blundell, J. B., & Jackson, G. A. D., in preparation
[25] Luckwill, L. C., & Whyte, P., *S.C.I. Monogr. 31*, 1968, p. 87 and personal communication.
[26] Jackson, G. A. D., & Blundell, J. B., *J. hort. Sci.*, 1963, **38,** 310
[27] Tincker, M. A. J., *Jl. R. hort. Soc.*, 1935, **60,** 399
[28] Jackson, G. A. D., & Blundell, J. B., *Nature, Lond.*, 1965, **205,** 518
[29] Jackson, G. A. D., Blundell, J. B., & Milborrow, B. V., in preparation
[30] Cornforth, J. W., Milborrow, B. V., & Ryback, G., *Nature, Lond.*, 1966, **210,** 627
[31] Milborrow, B. V., *Planta*, 1967, **76,** 93

[32] Jacobs, W. P., *A. Rev. Pl. Physiol.*, 1962, **13,** 403
[33] Rowley, G. D., *Am. Rose A.*, 1956, **41,** 70
[34] Pieniazek, J., & Rudnicki, R., *Bull. Acad. pol. Sci., Cl. II Ser. Sci. biol.*, 1967, **15,** 251
[35] Taylor, H. F., *S.C.I. Monogr. 31*, 1968, p. 22
[36] Jackson, G. A. D., & Blundell, J. B., *Nature, Lond.*, 1966, **212,** 1470
[37] Aspinall, D., Paleg, L. G., & Addicott, F. T., *Aust. J. biol. Sci.*, 1967, **20,** 869
[38] Thomson, P. A., *J. exp. Bot.*, 1961, **12,** 199
[39] Harrison, J. W. H., & Jackson, G. A. D., *Nature, Lond.*, 1945, **153,** 404
[40] Harrison, J. W. H., *Trans. nat. Hist. Soc. Northumb.*, 1920, **5,** 244
[41] Wright, S. T. C., *J. exp. Bot.*, 1966, **17,** 165
[42] Houghtaling, H. B., *Bull. Torrey bot. Club*, 1935, **62,** 243
[43] Paleg, L. G., *A. Rev. Pl. Physiol.*, 1965, **16,** 291
[44] Valdovinos, J. G., Ernest, L. C., & Henry, E. W., *Pl. Physiol., Lancaster*, 1967, **42,** 1803
[45] Heslop-Harrison, J., *A. Rev. Pl. Physiol.*, 1967, **18,** 325
[46] Crane, J. C., *Pl. Physiol., Lancaster*, 1965, **40,** 606
[47] Jacobs, W. P., & Case, D. B., *Science, N.Y.*, 1965, **148,** 1729
[48] Lang, A., & Nitson, J., *Ann. N.Y. Acad. Sci.*, 1967, **144,** 180
[49] Varner, J. E., *ibid.*, p. 219
[50] Key, J. L., Barnett, N. M., & Lin, C. Y., *ibid.*, p. 49
[51] Davies, D. D., Patterson, B. P., & Trewavas, A. J., *S.C.I. Monogr. 31*, 1968, p. 208
[52] Wareing, P. F., Good, J., Potter, H., & Pearson, A., *ibid.*, p. 191.
[53] van Overbeek, J., *ibid.*, p. 181
[54] Letham, D. S., *A. Rev. Pl. Physiol.*, 1967, **18,** 349

DISCUSSION

Dr. L. C. Luckwill.—With reference to the stimulation of parthenocarpy in *Rosa arvensis* by gibberellic acid, you mentioned that the growth of the hips was due solely to cell enlargement. This is in marked contrast to what we find in apple and pear. In those cultivars which respond to gibberellic acid the parthenocarpic fruits go through the normal phase of growth by cell division followed by cell expansion. Is *Rosa arvensis* unusual in this respect?

Dr. Jackson.—We have looked at *R. arvensis* and *R. sherardii*. In *R. arvensis* cell division continues for about 4 weeks after flowering, whereas in *R. sherardii* it is probably over much sooner. We treat them about one week before flowering when cell division in both species is active; in both cases however cell division stops after the application of gibberellin A_3 and growth is by cell enlargement only. These roses may be unusual in this respect; Bradley & Crane found similar results to yours in the peach. (Bradley, M. V., & Crane, J. C., *Bot. Gaz.*, 1962, **123,** 243).

Professor J. P. Nitsch.—Are the differences in the content of abscisic acid of different *Rosa* species related in any way to the levels of carotenoids in these species?

Dr. Jackson.—An interesting point, particularly in the light of what Dr. Taylor told us yesterday. The onset of intensive carotenoid production in the flesh and of dormancy in the embryo are roughly correlated, but we have no further information as to the levels of carotenoids in the various species.

HORMONE PHYSIOLOGY OF GROWTH REGULATION IN STRAWBERRY

By C. G. GUTTRIDGE*

(Scottish Horticultural Research Institute, Invergowrie, Dundee, Angus)

The effects of photoperiod, temperature and chilling on growth and flower formation in strawberry are briefly described, and the evidence for postulating a vegetative growth-promoting hormone is discussed. The hormone apparently inhibits flower formation. The similarity between the activities of the hormone and of exogenous gibberellins are reviewed and the tentative conclusion drawn that although increased biosynthesis of gibberellins is part of the photoperiodic response the natural hormone messenger between leaf and apex cannot be an active gibberellin like gibberellic acid, although it could be an inactive form or a precursor. There is evidence for a second independent photoperiodic factor operating on petiole growth. Photoperiodic stimuli and exogenous gibberellic acid are additive in their inhibition of flowering but this does not necessarily mean that the two act in the same metabolic pathway. Abscisic acid promotes flowering and extracts of dormant strawberry leaves contain abscisic acid. Extracts of non-dormant leaves and of stolon tips are active in the lettuce hypocotyl test for gibberellins. CCC (Chlormequat chloride; (2-chloroethyl)-trimethylammonium chloride) depresses vegetative growth but does not usually promote flower formation.

THE strawberry (*Fragaria* × *ananassa* Duch.) is a herbaceous perennial plant of temperate climates. It propagates by seed and by vegetative stolon offsets (horticulturally known as runners). The main stem or crown is short, giving a rosette sympodial habit, except when the stem elongates into an inflorescence.

The strawberry is a facultative SDP (short day plant).[1] After induction of flowering, inflorescences continue to grow to maturity but further inflorescences are not initiated unless the environment remains inductive. Prolonged exposure to short day, however, induces dormancy. Dormant plants are characterised by restricted, but not weak, vegetative growth and by a capacity to initiate inflorescences in environments which inhibit or delay flower formation in non-dormant plants.[2,3] Exposure to chilling temperatures releases the plants from dormancy and vegetative growth is then more vigorous and flower induction delayed (Table

* Present address: University of Bristol, Dept. of Agriculture and Horticulture, Long Ashton, Bristol.

Table I

Effect of daylength and chilling on the growth of strawberry plants (c.v. 'Talisman')

	Dormant		Chilled	
	SD*	LD*	SD	LD
Mean number of inflorescences per plant	2·33	0·13	1·53	0
Mean number of leaves to first inflorescence	1·1	>4·5	4·4	>7·5
Mean number of stolons per plant	0	2·53	3·40	6·07
Mean length of petioles mm (a)	32	72	75	94

*The plants, in pots, were transferred from outdoors on either 17th December (Dormant) or 25th February (Chilled) into one of two standard environments in growth rooms and grown there for 12 weeks. There were 15 plants per treatment. The two environments were short day (SD) consisting of 10 h lighting with fluorescent and incandescent lamps in the wattage ratio of 6:1 at an intensity of approximately 900 f.c. and long day (LD), consisting of the same basic illumination but with the incandescent lighting extended to 18 h. The temperature was 15·5°c. The chilled plants were overwintered in well-ventilated garden frames, exposed to natural chilling temperatures.

(a) Mean of first 4 leaves to emerge on the principal crown after treatments commenced.

I). Long daylengths promote vegetative growth, large leaves, long petioles and the formation of stolons[1] in axillary sites. In short day, petioles and leaves are small, rosetted branches replace stolons and inflorescences are initiated terminally on main stems and branches. Light breaks with incandescent light at intensities of one or more foot candles effectively substitute for long daylengths.[4,5] The promotive effect of long day on vegetative growth is lessened by low temperature, mineral starvation or drought and in any of these circumstances flower formation may not be inhibited by long day.[1]

There is an antagonism between flowering (by which is meant initiation of the inflorescence) and vegetative growth. Changes in the environment which promote vegetative growth decrease the likelihood of flowering, and flowering is inhibited if thresholds are attained. This antagonistic relationship holds for changes in photoperiod,[4] temperature[1,6] (except perhaps below about 10°), and for the effects of chilling.[2]

Experiments with donor/receptor units formed by mother and daughter plants joined by the stolon, have shown that exposure of the donor plant to long day stimulates the production of a hormone which is transported along the stolon to the receptor plant in short days where it induces increased vegetative development and inhibits flower formation.[7-9] Experiments of this kind have consistently given the same result and contrary evidence has not been found.[3,10] The term hormone seems appropriate in this situation first because the stimulus is demonstrably transmissible and, secondly because it arises from illumination with low-intensity lighting and is thus unlikely to be solely photosynthate. However, it is not clear whether the long day hormone is a single or a multiple entity or whether it inhibits flowering directly at the apex or indirectly as a consequence of promoted vegetative growth. One possibility is that flowering fails in long day because the production of a florigenic hormone is inhibited. However, several attempts with donor/receptor units to demonstrate the existence of a florigenic hormone have failed.[8-10] Neither promotion of flowering nor suppression of vegetative growth attributable to the activity of a vegetative growth inhibitor or a florigenic hormone has been detected in receptor plants in long daylengths. Flowering only occurred in receptor plants when they were defoliated. Some results of Hartmann[11] purport to demonstrate the existence of a florigenic hormone arising in short day in strawberry, but his experiment is open to criticism. Control plants were not comparable and the very real possibility exists that induction occurred before the receptor plants were perceptive of the long day. The lighting used was deficient in both red and far red radiations, so that a fully effective long day was not achieved.

Although experiments have yielded no evidence for the existence in strawberry of either a vegetative growth inhibitor or a florigenic hormone there is evidence (1) that there is a minimum leaf requirement[12] but this does not necessarily imply that leaves produce a special florigenic factor, or one that has a regulatory role in flowering, and (2) that abscisic acid may in fact function as a florigenic agent in strawberry although it may do so only in marginal situations. Abscisic acid has been found in extracts of dormant strawberry leaves by Gabr & Guttridge,[13] and independently by Rudnicki, Pieniazek & Pieniazek.[14] El-Antably, Wareing & Hillman[15] have shown that synthetic abscisic acid induces flower formation in strawberry and in some circumstances inhibits stolon formation. Analogy with other species suggests that abscisic acid may induce dormancy in strawberry. Because of these results with abscisic acid it is surprising that no evidence for a flower-promoting or a growth-inhibiting hormone was found in the environmental experiments. The discovery of abscisic acid is too recent to know how to interpret this situation, but one possible explanation is that native abscisic acid is not formed in mature leaves. It would not then be likely to be transmitted along the stolon. A second possibility is that the inhibitor is relatively easily antagonised by long day factors.

Much evidence suggests that the long day hormone in strawberry is a gibberellin or at least that gibberellin mediates in the response to long daylengths. When gibberellic acid is applied to strawberry plants in short day, petioles and leaves and existing inflorescences increase in size, stolons are induced to form in axillary sites and flower formation is inhibited, all in a manner resembling the effects of increasing the photoperiod.[16,17] In some situations flower formation is inhibited by low doses of exogenous gibberellic acid. Table II shows the results of an experiment in which flower initiation was delayed by a dose of less than 0·1 μg GA_3 per gramme dry wt. of plant tissue per day and inhibited by

Table II

Effect of application of gibberellic acid (GA_3) on flower formation in strawberry (c.v. 'Talisman')

Mean No. of inflorescences initiated per plant during the first four plastochrons from the start of treatment.

GA_3 dose (a)	Plastochron No. 1	2	3	4
0·0	0·13(b)	0·31	0·31	0·25
0·087	0·0	0·0	0·37	0·25
0·261	0·0	0·0	0·0	0·31
0·870	0·0	0·0	0·07	0·0

The plants were raised from runners outdoors in autumn in 5 in. pots. During treatment, which commenced on 10th December, they were grown in a growth cabinet at a temperature of 19°c, illuminated daily for 11 h with warm white fluorescent lamps and incandescent lamps in the wattage ratio of 6·1 at an intensity approximately 1000 f.c. There were 16 plants per treatment.

(a) Dose of GA_3 per day in μg/g dry weight of plant tissue at the start of treatment. The GA_3 was applied on alternate days at the actual rate of 0·15, 0·45 or 1·5 μg/plant in a microdrop of 0·002, 0·006 or 0·02 ml respectively. The plastochron was approximately 8 days.

(b) The maximum theoretical mean flowering response per plastochron is 0·5 but is more usually 0·33 under optimum conditions for flowering (i.e. an inflorescence after every third node).

10 times this amount. These doses are similar to those needed to elicit a response in petioles. In marginal situations the native or endogenous photoperiodic stimulus and exogenous gibberellic acid are additive in inhibiting flowering. This is shown in an experiment in which strawberry plants were grown in photoperiods increasing from 12 to 15 h by ½ h steps and received weekly sprays of water or aqueous solutions of gibberellic acid (GA_3) containing 40, 70 or 100 ppm. The results (Fig. 1) show that the long day hormone and

Gibberellic acid	Photoperiod (DL)						
	11+1	11+1·5	11+2	11+2·5	11+3	11+3·5	11+4
0	●●● ●●● ●●● ●●● ●●●	●●● ●●● ●●● ●●● ●●●	●●● ●●● ●●● ●●● ●●●	●●● ●● ●● ● ●	●●● ●● ●● ●●	●● ●	●●● ●● ●
40	●●● ●●● ●●● ●●● ●	●●● ●● ●● ● ●	●● ●● ●● ●	● ● ●	●	●	●
70	●● ●● ●	●●● ●● ●● ●● ●	●●● ●● ●● ● ●	●		●●	
100	● ● ● ●		● ● ●		●	●	

FIG. 1. *Interaction of photoperiod and exogenous gibberellic acid on flowering in Strawberry*

Plants of the cv. 'Talisman' were grown in growth cabinets at a temperature of 19°C and illuminated daily for 11 h with warm white fluorescent lamps and incandescent lamps in the wattage ratio of 6:1 and an intensity of approximately 1500 f.c. The photoperiod was extended beyond 11 h by supplementary lighting with incandescent lamps alone to give photoperiods of 11+1, 11+1½, 11+2, 11+2½, 11+3, 11+3½, and 11+4 h. The plants were sprayed on the 1st, 8th and 15th day of treatment with gibberellic acid (GA_3) at concentrations of 40, 70 or 100 ppm with 0·5% Shellestol or with 0·5% Shellestol alone (controls). After 24 days in the cabinets the plants were returned to LD in the glasshouse. Three crowns from each of the five plants per treatment were dissected and the presence of an inflorescence recorded.

● indicates presence of an inflorescence. Vegetative crowns are represented as blanks.

exogenous gibberellic acid supplement each other when either is present in marginal amounts. Saturating doses of either will completely inhibit flowering but such levels were not attained in this experiment. This result does not prove rigorously that exogenous gibberellin is capable of replacing the endogenous hormone in a chemical reaction, but it could be so.

Such replacement effects in a natural system are more convincingly demonstrated in kinetic studies[18] in which the level of one factor can be shown to have a diminishing effect on growth rate with increasing levels of a second factor and no effect when the second factor is saturating. In strawberry the rate of petiole growth has been found to be proportional to the final petiole length when growth increments are induced either by photoperiod or by exogenous gibberellic acid.

The results of such an experiment show (Fig. 2) that saturating doses of exogenous gibberellic acid replace a part, but only a part, of the long day stimulus. It is apparent from this and similar experiments that long photoperiods induce the formation of a second and distinct kind of factor, independent of that which appears to be replaced by exogenous gibberellic acid.

The replacement of the endogenous factor by exogenous gibberellic acid and the close parallel between the responses of the plant to long day and to exogenous gibberellic acid make it seem likely that exposure to long day increases the biosynthesis of native gibberellins. However, a gibberellin such as GA_3 may not be present in the assimilate stream which emanates from leaves in long daylengths. This is because exogenous gibberellins cause the vegetative stem to elongate (caulesce)[16] in the same manner in which stems of some biennials are caused to elongate by gibberellins. Strawberry is a clear case of gibberellins causing stem elongation in a rosette plant without flower formation.

If the caulescence of vegetative stems could be accounted for there would clearly be a case for considering whether one of the long-day hormones is a gibberellin. One possible explanation for the absence of stem caulescence is that the native hormone in the assimilate stream, supposing that it is an active gibberellin, is differentially distributed and does not reach the sub-apical region; a second possibility is that it reaches the region but is inactivated there. However, GA_3 applied to the foliage experimentally reaches the sub-apical region and is not inactivated. Caulescence might result if the differential distribution or inactivation processes were being rendered ineffective or swamped by exogenous doses greatly in excess of natural physiological levels. Although there is no

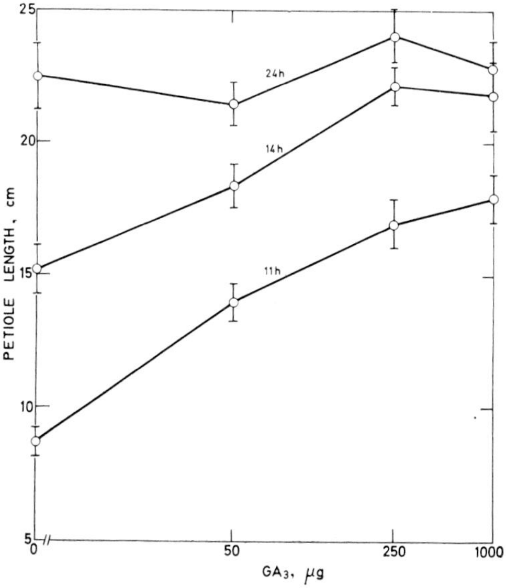

FIG. 2. *Effect of daylength and gibberellic acid on strawberry petioles*

The plants (cv. 'Redgauntlet'), established in pots, were transferred from outdoors on 16th February and treated with gibberellic acid (GA_3) by the uptake of 0·5 ml of solution into the cut surface of a delaminated petiole. The solutions of GA_3 were made up in NaOH to give doses of 0 (control), 50, 250 and 1000 µg per plant. The rate of uptake of the solutions varied and in some instances plants were excluded from the experiment if uptake was unsatisfactory. After treatment the plants were grown in growth cabinets at 16° in daylengths of 11, 11+3 or 11+13 h duration (11 h illumination with warm white fluorescent lamps and incandescent lamps in the wattage ratio of 6:1 at an intensity of approximately 1000 f.c., supplemented by 0, 3 or 13 h with lighting from incandescent lamps only.) Petiole lengths are means of the first, second and third leaves to emerge after treatment commenced. There were 14 plants per treatment. Standard errors (SEM) are indicated.

rigorous evidence on the point, experimental results suggest that caulescence is caused by doses that are within the physiological range. In the first place caulescence is induced by single doses of the order of 0·5 µg per gram dry weight. Secondly and perhaps more convincingly the results in Fig. 3 and Table III show clearly that doses of GA_3 which induce responses on petioles within the range of those induced by photoperiod also induce stem caulescence, which photoperiod does not.

In species in which flowering is not inhibited by applied gibberellic acid, 0·5 µg GA_3 per plant caused a few transversely orientated cell divisions in *Samolus*[19] (probably a much more sensitive criterion of elongation than macroscopically measurable elongation in strawberry) and 2 μg/plant causes caulescence in *Centaurium minus*.[20] In neither case were plant weights reported, but the dosages are of the same order as those inducing a response in strawberry.

There remains therefore, the hypothesis that the native hormone is probably not GA_1, GA_3, GA_4, GA_7, GA_9[21] or GA_{13} all of which induce caulescence (assuming the absence of impurities in the samples). Perhaps the hormone mediating between leaf and growing tissues is a gibberellin precursor or an inactive form of gibberellin or even an unrelated substance which unlocks gibberellin biosynthesis. Three possible precursors of gibberellin, kaurene, kaur-16-en-19-oic acid and kaur-16-en-19-ol are not active on

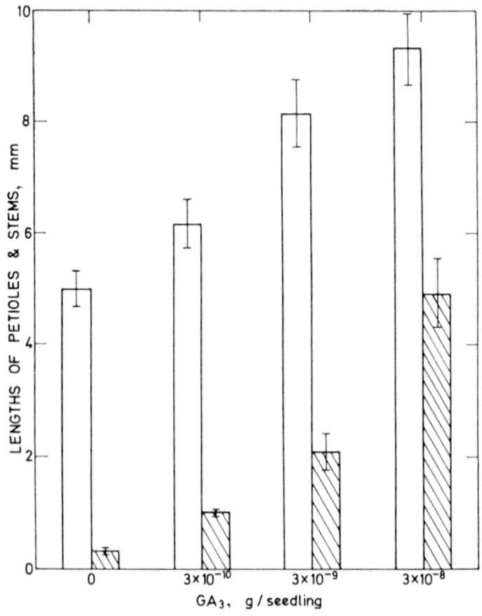

FIG. 3. *Effects of gibberellic acid on length of stems and petioles of seedlings of* F. vesca semperflorens

Seedlings of *Fragaria vesca semperflorens* cv. 'Brillante', with cotyledons fully expanded, were grown in four different light regimes in growth cabinets. Gibberellic acid (GA_3) was applied in single microdrops of 0·003 ml to the bud of each seedling in concentrations of 0 (control), 0·1, 1·0 or 10 μg/ml. in 30% ethanol in water. The results for the four light regimes were pooled for the purpose of this diagram. Standard errors (±) are indicated.

Petioles—plain histograms Stems—shaded histograms

Table III

Effect of exogenous gibberellic acid on petiole length and stem length
(cv. 'Talisman')

GA_3 dose per plastochron, $\mu g/g$ dry wt.	Day length	Petiole lengths, % control	Stem length, % control
0	SD	100	100
0·7	SD	115	120
2·1	SD	136	157
0	LD	201	108

The plants were grown in growth cabinets in an 11 h (SD) photoperiod (at a light intensity of approximately 1000 f.c. of warm white fluorescent and incandescent lighting in the approximate wattage ratio of 6:1). Gibberellic acid was applied in microdrops to the foliage, four times per plastochron. The results (means of 64 plants) are expressed as percentages of controls. The plants in long day (LD) received supplementary lighting with incandescent lamps for 6 h daily, following the main light period. Dosages of gibberellic acid are expressed in μg GA_3 per g dry weight of a sample of plants weighed at the beginning of the experiment.

petioles (or stems) when applied at 100, 500 or 1000 times the concentration of GA_3 that is required for a clear response, so it seems unlikely that any of these three compounds is the hormone messenger. This argument does not exclude the possibility that gibberellins already identified above occur in strawberry, but the evidence suggests that the native hormone in the assimilate stream cannot be one of them.

Extracts of strawberry runner tips and of young leaves are active in the lettuce hypocotyl test[22] for gibberellins (P. B. Goodwin, & A. Gordon, unpublished results) and similar demonstrations of gibberellin-like activity have been reported,[23,24] but no gibberellins have yet been positively identified.

The mode of action of exogenous gibberellins when they inhibit flowering is not understood. When other species are considered it is found that the inhibition of flowering by gibberellic acid cuts across the photoperiodic division of plants into long day and short day species although all species from which effects have been reported are perennials. For example, flowering is inhibited by exogenous GA_3 in the SDP's, *Kalanchoe blossfeldiana*,[25] *Zygocactus truncatus*,[26] *Euphorbia pullcherima*,[27] and likewise inhibited in the LDP's (long day plants) *Prosperpinacea palustris*[28] and *Fuchsia* spp.[29] The full list includes several woody species and altogether 14 genera are represented. In contrast, in many LDP's and in biennials, GA_3 induces flowering, although it is not generally thought to be the flowering hormone. The mode of action of GA_3 as a flower inhibitor may not be the same in all the species. Whether or not gibberellins replace native hormones in a precise chemical sense is difficult to establish. Lockhart's principles[30] cannot be applied to processes which lead to the inhibition of single, non-quantitative events like flower initiation where saturating levels of either factor completely inhibit flowering, as in strawberry.

The antagonistic correlation between flowering and vegetative growth occurs in other species besides strawberry and prompts the questions whether the hormone or gibberellic acid inhibit flowering directly, perhaps at the apex, or whether they promote vegetative growth only and inhibit flowering indirectly, perhaps by diverting photosynthates from the apex. The diversion hypothesis seems unlikely because various burdens, such as carrying a heavy fruit crop or a large number of runner plants, actually promote rather than inhibit flower formation. It seems more likely that the diversion of photosynthates promotes flowering.

The correlation between flowering and vegetative growth in strawberry is broken in three circumstances. First, when the growth retardant CCC is applied, vegetative growth[31] is depressed but flowering is not promoted. No promotion has been observed in numerous experiments (for example, see Table IV), although a

Table IV

Effect of CCC on flower formation in strawberry (cv. 'Talisman')

Daylength	CCC, g/plant	Number of nodes to first inflorescence	Petiole length, cm
SD	0	2·14	7·2
	0·03	2·57	7·4
	0·25	2·29	7·0
LD	0	6·75	14·2
	0·03	6·63	12·3
	0·25	7·75	10·1

The plants were grown in horticultural compost in 4 in. dia. pots in growth cabinets at a temperature of 16°c. Lighting was provided by warm white fluorescent and incandescent lamps in the wattage ratio of approximately 6:1 at an intensity of approximately 900 f.c. The photoperiod was either 11 h (SD) or 11+7 (LD) in which the supplementary 7 h was supplied by incandescent lamps only. CCC was applied to the soil at the rates given, in 50 mls of water at the beginning of the experiment.

slight promotion of flowering has been reported.[15] This effect is difficult to interpret because of evidence from other species that CCC suppresses gibberellin biosynthesis, an action which if it occurred in strawberry might be expected to promote flowering. Possibly a florigenic factor is required in strawberry, and this might be abscisic acid. Secondly, in the everbearing strawberries vegetative growth is promoted by long day, but flowering is not inhibited. However, when vegetative growth is promoted by application of gibberellic acid, flowering is inhibited. It is difficult to see how this could occur if flower inhibition resulted merely from starvation of the apex. Thirdly, there is some evidence from *Fuchsia*[32] and from strawberry[21] that exogenous gibberellic acid inhibits flowering more efficiently when applied near the apex than on the foliage. Whilst these pieces of circumstantial evidence do not exclude the possibility that flowering is controlled by the level of vegetative growth *per se*, a hormonal regulation seems a more plausible alternative and it may be that the native hormones act directly at the apex.

Clearly there are many unanswered questions, and the hormone regulation of growth in strawberry will not be understood until more is known about (a) the role of abscisic acid and (b) the nature of the second factor, which apparently is independent of the gibberellin system for petiole elongation. Progress with these problems at the physiological level is possible with present tech-qiques, but the chemical identification of the hormone messenger and of the second factor must also be investigated.

Acknowledgements

The author gratefully acknowledges the help and comments of his colleagues Dr. P. A. Thompson, Dr. P. B. Goodwin and Mr. A. Gordon, some of whose unpublished data have been included in this paper.

References

[1] Darrow, G. M., & Waldo, G. F., *Tech. Bull. 453 U.S. Dep. Agric.*, 1934
[2] Guttridge, C. G., *J. hort. Sci.*, 1958, **33,** 119
[3] Jahn, O. L., & Dana, M. N., *Proc. Am. Soc. hort. Sci.*, 1966, **89,** 322
[4] Borthwick, H. A., & Parker, M. W., *Proc. 13th Int. Hort. Congress*, 1953, **2,** 801
[5] Roodenburg, J. W. M., *Proc. 8th Int. bot. Congress, Paris*, Sect. 11 & 12, 1956, p. 318
[6] Went, F. W., 'Chronica Botanica', 1957, **13,** Ch. 9 (Waltham, Mass.)
[7] Guttridge, C. G., *Nature, Lond.*, 1956, **178,** 50
[8] Guttridge, C. G., *Ann. Bot.*, 1959, **23,** 351
[9] Guttridge, C. G., *Ann. Bot.*, 1959, **23,** 612

[10] Jonkers, H., *Meded. 65 LandbHoogesch. Wageningen*, 1965
[11] Hartmann, H. T., *Pl. Physiol., Lancaster*, 1947, **22**, 407
[12] Mason, D. T., *Hort. Res.*, 1967, **7**, 97
[13] Gabr, O. M. K., & Guttridge, C. G., *Planta*, 1968, **78**, 305
[14] Rudnicki, J., Pieniazek, J., & Pieniazek, N., *Bull. Acad. pol. Sci. Cl. II Ser. Sci. biol.*, 1968, **16**, (in press)
[15] El-Antably, H. M. M., Wareing, P. F., & Hillman, J., *Planta*, 1967, **73**, 74
[16] Thompson, P. A., & Guttridge, C. G., *Nature, Lond.*, 1959, **184**, B.A.72
[17] Porlingis, I. C., & Boynton, D., *Proc. Am. Soc. hort. Sci.*, 1961, **78**, 261
[18] Lockhart, J. A., *Am. J. Bot.*, 1961, **48**, 516
[19] Sachs, R. M., Bretz, C. F., & Lang, A., *Am. J. Bot.*, 1959, **46**, 376
[20] McComb, A. J., *Planta*, 1967, **76**, 242
[21] Guttridge, C. G., & Thompson, P. A., *J. exp. Bot.*, 1964, **15**, 631
[22] Thompson, P. A., *A. Rep. Scot. hort. Res. Inst.*, 1964, **11**, 50
[23] Porlingis, I. C., & Boynton, D., *Proc. Am. Soc. hort. Sci.*, 1961, **78**, 256
[24] Leshem, Y., & Koller, D., *Ann. Bot.*, 1966, **30**, 587
[25] Harder, R., & Bunsow, R., *Planta*, 1958, **51**, 201
[26] Fujihara, K., *J. hort. Ass. Japan*, 1959, **28**, 200
[27] Guttridge, C. G., *Nature, Lond.*, 1963, **197**, 920
[28] Davis, G. J., *Pl. Physiol., Lancaster*, 1967, **42**, 667
[29] Sachs, R. M., & Bretz, C. F., *Proc. Am. Soc. hort. Sci.*, 1962, **80**, 581
[30] Lockhart, J., *A. Rev. Pl. Physiol.*, 1965, **16**, 37
[31] Guttridge, C. G., *Hort. Res.*, 1964, **3**, 79
[32] Sachs, R. M., Kofranek, A. M., & Shiow-Ying Shyr, *Am. J. Bot.*, 1967, **54**, 921

DISCUSSION

Dr. J. Bruinsma.—The parent-daughter experiments failed to show a short day factor. The results of the long night interruptions, however, are strongly suggestive of its existence. Assuming that abscisic acid could be involved, it is unlikely that mature tissue should be unable to synthesise it. Would you comment on the suggestion that abscisic acid is readily destroyed in long day plants and is thereby prevented from exerting its influence?

Dr. Guttridge.—I think your suggestion is a possibility that cannot yet be excluded. But if it is correct it would not then be immediately clear why dormant plants exhibit a relatively depressed growth habit for several weeks even in LD.

Professor B. B. Stowe.—Sironval has reported that lipid extracts, and in particular vitamin E, promote flowering in strawberries. Since this substance has isoprenoid similarities to abscisic acid could there be an analogy to your experiments?

Dr. Guttridge.—Sironval's results do not seem to have any relevance here. He found that when immature stolons were isolated from the mother plant, existing inflorescences on a proportion of them reverted to vegetative growth. Application of lipid extracts and of vitamin E decreased the proportion which reverted. As I understand the situation, the formation of new inflorescences was not promoted either by extracted lipids or by vitamin E in any experiments. Moreover no regulatory role could be ascribed to these compounds as a result of these experiments, because it appears that flowering in the species with which Sironval worked is not subject to environmental control except to the extent that increased flowering is associated with improved growth in LD compared with that achieved in SD.

EFFECTS OF PLANT GROWTH REGULATORS ON THE FLOWERING OF *PHARBITIS NIL*

By HIROSHI HARADA

(*Groupe de Physiologie Pluricellulaire, Le Phytotron, C.N.R.S., Gif-sur-Yvette, France*)

Effects of morphactin, gibberellin, indoleacetic acid and kinetin on the flowering of young *Pharbitis* plants were studied. Simultaneous application of morphactin and gibberellin A_3 showed synergistic inhibitory action on flowering, and also modified the number and the position of flower buds on the main stem. Morphactin retarded the flowering of terminal flowers and accelerated that of lateral flower buds. This phenomenon could be reversed partially or completely by the application of indoleacetic acid or gibberellin A_3, whereas kinetin was ineffective.

Introduction

THE physiology of flowering is one of the most exciting problems in plant physiology. Many elaborate experiments on flowering have been undertaken using numerous different kinds of wild as well as cultivated plants. Among them, 'cocklebur', *Xanthium pennsylvanicum,* has been most widely adopted for research on flowering. A physiological feature of *Xanthium* is its ability to initiate flower primordia in response to a single short day. However, this species must be 1 to 2 months old, before acquiring the highest sensitivity to photoperiod.

Japanese morning glory, *Pharbitis nil,* is another short-day plant which requires a single short day to bring about high flowering response. In contrast to *Xanthium, Pharbitis nil* becomes very sensitive to photoperiod only 2 to 3 days after germination. It is therefore highly suitable material for research on flowering and has been often used by many investigators, initially in Japan and recently also in the Western World.

Pharbitis plants remain strictly vegetative for an extended period under long days or continuous illumination at 20° or above. Nakayama[1] and Ilkeda[2] reported that the optimum night temperature for floral induction in *Pharbitis nil* is between 25° and 30° and that temperatures above 32° are inhibitory. However, optimal temperature for flowering has never been precisely determined.

Ikeda[2] also reported that if plants were exposed to very high temperatures, for example 34°, after the inductive dark period, flowering was reduced to some extent, but low temperatures, for example 10°, had less effect. Sensitivity to photoperiod varies from strain to strain. A strain called 'Violet' is one of the most sensitive[3] and most widely employed for studies on flower physiology.

Before presenting experimental results, brief mention should be made of what is known about the effects of various growth substances on the flowering of *Pharbitis nil.* It is well known that flower initiation in short-day plants is inhibited by auxin applied just before or during the inductive dark period. This also occurs with *Pharbitis nil.* According to Ogawa,[4] flowering was inhibited when indole-3-acetic acid was applied before or during the first half of the dark period. Indoleacetic acid applied to the cotyledons was more effective than that applied to the plumules. These results suggest that indoleacetic acid inhibits some parts of a biochemical process taking place in the cotyledons, rather than the morphogenetic reactions in the apex.

Ogawa & Imamura[5] reported also that the flowering response of *Pharbitis nil* was increased by gibberellin A_3 if plants were treated before, during, or immediately after a short inductive dark period. Application to cotyledons was again more effective than that to shoot apices.

Ogawa[6] reported the promotive effect on flowering of kinetins applied to cotyledons before an inductive dark treatment. He showed that as little as 100 μg/l promoted flowering, but maximum promotion was obtained at concentrations ranging between 50 and 100 mg/l.

With regard to the so-called growth retardants or dwarfing compounds, Zeevaart[7] showed that all of four compounds he employed inhibited flower formation, 'Phosfon D' (tributyl-2,4-dichlorobenzylphosphonium chloride) being most effective, followed by AMO-1618 (2-isopropyl-4-dimethylamino-5-methylphenyl 1-piperidinecarboxylate, methyl chloride), CCC (chlormequat chloride; (2-chloroethyl)-trimethylammonium chloride) and B-995 (*N*- dimethylaminosuccinamic acid) in this order. He obtained the strongest suppression of flowering with applications immediately before or during the inductive dark period. The inhibitory effect of CCC was overcome by application of gibberellin A_3. On the other hand, auxins were completely inactive in reversing the inhibitory effects of B-995. Under non-inductive long-day conditions CCC was slightly promotive, inducing some flower formation, an effect that was prevented by the application of gibberellin A_3.

Recently, El-Antably & Wareing[8] showed some promotive

effect on flower initiation of abscisic acid, which belongs to another group of plant growth regulators.

Still another group of plant growth regulators are collectively termed 'morphactins' (M) because of their action on plant morphogenesis. Morphactins are a group of fluorene compounds synthesised in the research laboratories of E. Merck AG., Darmstadt, Germany. Interesting effects of these compounds on plant morphogenesis was first noted by the Merck research group in 1960.[9] Some of their typical effects include inhibition of development, particularly inhibition of elongation of internodes, production of deformed growth and fasciation, disturbance of geotropic and phototropic responses, induction of parthenocarpic fruit development, and stimulation of branching in stems and roots.

Interactions between morphactins and growth substances, especially gibberellin A_3, have been reported by several workers. Tognoni and co-workers[10] have reported that morphactins were without effect in reducing the response of dwarf peas to gibberellin, when applied either prior to or simultaneously with gibberellin A_3. Some morphactins in combination with gibberellin A_3 produced taller plants than the gibberellin-treated controls. The same authors and also other workers[11,12] showed that morphactins did not have any effect on α-amylase biosynthesis in embryo-less barley half-seeds, either with or without simultaneous application of gibberellin, and suggested that morphactins do not compete with gibberellin for similar sites of action. Experimenting with seedlings of several plant species, Khan[13] reported that neither indoleacetic acid nor gibberellin A_3 could restore lost geotropic response caused by morphactin treatment, Working with *Begonia* leaf discs, Schott & Schrandolf[14] showed that gibberellin-induced inhibition of adventitious root and shoot formation, and increase in area of discs, were not affected by morphactin treatment, although morphactin itself increases the number of regenerating shoots. In the case of seed germination, gibberellin can reverse partially or completely the morphactin-induced inhibition in lettuce[15] and *Primula obconica*.[16] According to Sankhla,[17] kinetin is much more active than gibberellin in overcoming the inhibitory effect of morphactin in the lettuce seed germination.

Much less is known about the effect of morphactins, alone or in combination with other growth regulators on flowering, than is known about their effect on other phases of plant growth. The present investigation was therefore undertaken to study the interactions between morphactin and growth substances on flowering of *Pharbitis nil*.

Materials and methods

'Violet', one of the most photoperiodically sensitive strains of *Pharbitis nil,* shown in Fig. 1, was employed throughout the present study. To obtain uniform germination, seeds were immersed in concentrated sulphuric acid and were stirred occasionally for 1 hour, and were thoroughly washed in running water overnight. The next morning, the swollen seeds were placed, hilum up, in 8 cm pots filled with vermiculite and glass-wool, and kept at 27° under artificial light of about 14,000 lux. Six days after the H_2SO_4 treatment, the seedlings with fully developed cotyledons were selected for uniformity and were subjected to different treatments.

The morphactin used was methyl-2-chloro-9-hydroxyfluorene-(9)-carboxylate, and the growth substances employed in combination with the morphactin were indole-3-acetic acid (IAA), gibberellin A_3 (GA_3) and kinetin (K). For each growth substance only one concentration which generally gives clear effect on flowering of *Pharbitis nil* was selected, namely, 100 mg/l IAA and GA_3, and 50 mg/l K. The morphactin in concentrations of 20 mg/l and 1 mg/l was applied to leaves and roots, respectively. All

FIG. 1. *Flowering* Pharbitis nil, *strain 'Violet'*
The plant has been grown under continuous short days of 9 hours from the 7th day after the seed treatment with H_2SO_4

solutions, except that for root treatment were 0·1% with respect to 'Tween 80'. In the case of leaf treatments the solutions were applied to both adaxial and abaxial surfaces of cotyledons by means of small camel-hair brushes. The three growth substances were also applied to the plumules. Root applications were made by incorporating 10 ml of the morphactin solution into the culture medium. Generally 10 uniform plants were selected for each treatment and each experiment was repeated twice. All experiments were carried out in the controlled light rooms of the Phytotron with light intensity approximately 14,000 lux at plant level. Long-day conditions consisted of 16 hours of light and 8 hours of dark, and short-day conditions consisted of 9 hours of light and 15 hours of dark.

The following criteria were adopted to evaluate the effects of various treatments on flowering: (i) the percentage of plants which initiated at least one flower bud, (ii) the percentage of plants which set terminal flower buds (young *Pharbitis* plants usually develop terminal flower buds only when sufficiently induced), (iii) the number of flower buds on the main stem, (iv) the positions (node number) of flower buds on the main stem, (v) the order of opening of flower buds, and (vi) number of days required to the opening of the first flower from the beginning of the inductive treatments.

Results and discussion

Effects of temperature

This experiment was carried out to demonstrate how sensitive young *Pharbitis* seedlings are to temperature in regard to flowering. The seedlings were first divided into 2 groups and each group was given a single inductive dark period under two different temperature conditions. One group was induced at a constant temperature of 27° (condition I), and the other group was placed in the dark at 32°, while the temperature was gradually lowered to 27° during the first 1½ hours (condition II). In other words, the second group was induced at a little higher temperature than the first group only for 1½ hours at the onset of 15 hours darkness. The results obtained are summarised in Table I.

The amount of flowering stimulus produced during a single dark period was greater under condition II (see Table I) than under condition I, because all plants induced under condition II set a terminal flower, regardless of the subsequent temperature. In the case of the plants induced at 27° (condition I), the expression of flowering stimulus became weaker as the temperature after induction increased. Therefore, the temperature of 27° during the inductive dark period is certainly not high enough for maximum induction. On the other hand, the temperature after induction must be kept fairly low for the maximum expression of floral stimulus. Slight decrease in number of flower buds per plant with decreasing temperature after induction is mainly due to the lower position of the terminal bud which, in turn, indicates a higher degree of floral induction.

Table I

Effects of temperature during and after an inductive dark period on flowering of Pharbitis nil.

Experimental conditions (see text)	Single 15 h dark period given at	After induction, grown under long days at	Plants with floral apex, %	Plants with vegetative apex, %	Average number of flower buds per plant
Ia	27° constant	22°	100	0	6·5±0·26
Ib		24°	70	30	7·5±0·68
Ic		27°	20	80	4·3±0·63
Id		27°—32°*	10	90	1·6±0·26
IIa	32° initially, then lowered to 27°	22°	100	0	5·1±0·27
IIb		24°	100	0	6·5±0·26
IIc		27°	100	0	6·8±0·24
IId		27°—32°	100	0	7·0±0·36

*27° for 15 h and 32° for 9 h.

Effects on inhibition of flowering

The experimental conditions were the same as II-d in Table I. The morphactin was applied to one of the paired cotyledons and the growth substances were applied to the opposite cotyledons as well as to the plumules, before and after a single inductive dark period. As shown in Table II, the treatments with indoleacetic acid alone, and indoleacetic acid + morphactin completely inhibited flowering, while the treatments with gibberellin A_3, kinetin, morphactin, and morphactin + kinetin did not. Interesting results were obtained by simultaneous application of the morphactin and gibberellin A_3. This treatment caused partial inhibition of flowering ,and both methods of the morphactin application, either through cotyledons or through roots, were effective. As far as can be ascertained, this is the first demonstration of the synergistic inhibitory effect between the morphactin and gibberellin on flower formation. Histochemical experiments are needed in order to see what part of the flowering process is disturbed by these two substances, and these are in progress.

Table II

Effects of four plant growth regulators on flowering of Pharbitis nil

	Flowering plants, %	Days to first flowering	Plant height cm
Control	100	25	83
IAA 100 mg/l	0	—	—
GA 100 mg/l	100	23	95
Kinetin 50 mg/l	100	23	91
M 20 mg/l (cotyledon)	100	26	77
M 20 mg/l +IAA 100 mg/l	0	—	—
M 20 mg/l +GA 100 mg/l	75	(25)	—
M 20 mg/l + K 50 mg/l	100	26	81
M 1 mg/l (root)	100	25	65
M 1 mg/l + IAA 100 mg/l	25	(26)	—
M 1 mg/l + GA 100 mg/l	50	(25)	—
M 1 mg/l + K 50 mg/l	100	26	78

Figures in parentheses indicate the averages of the flowering plants only.

Effects on position of flower buds

This experiment shows another example of morphactin–gibberellin interaction on the flowering of *Pharbitis nil.* Six days after seed treatment with H_2SO_4, the seedlings were given five successive short days at 22°, then they were placed under long days at 27°. The seedlings were treated four times with the growth regulators, namely once just before the first long night and three times during the short-day treatment. Each of the plants in the different treatments gave a terminal flower, and all nodes between the lowest flowers and the terminal flowers set lateral flower buds. However, as shown in Table III, the positions of the lateral and the terminal flowers, and subsequently the total number of flowers were modified depending on treatments. For example, morphactin alone applied through roots had no effect, giving a vegetative bud at the first (cotyledonary) node only. On the other hand, the plants treated with gibberellin A_3 gave vegetative buds on the first and the second nodes, and a terminal flower at the seventh position. Furthermore, the plants treated with gibberellin A_3+morphactin through roots gave vegetative buds on the first, second and third nodes, and the first flower only at the fourth node, with the terminal flower at the sixth position. Without further experiments, it is difficult to interpret these actions of the growth regulators.

However, there are at least two possible explanations: firstly, these substances may modify the movement or the distribution of the flowering stimulus in some way; secondly, they might modify the sensitivity of lateral buds to the flowering stimulus, rendering lower buds less sensitive and higher buds more sensitive.

Table III

Effect of morphactin (M) and gibberellin A_3 (GA_3) on the position and the number of flower buds on the main stem of Pharbitis nil

	Node number of main stem								Total number of flower buds
	1st*	2nd	3rd	4th	5th	6th	7th	8th	
Control	X	O	O	O	O	O	O	T	7
M	X	O	O	O	O	O	O	T	7
GA_3	X	X	O	O	O	O	T		5
$M+GA_3$	X	X	X	O	O	T			3

*: cotyledonary node
X: vegetative bud.
O: flower bud.
T: terminal flower.

Effects on flowering order

This experiment dealt with the effects of the morphactin and the growth substances on the order of flowering of young *Pharbitis* plants. The experiment was carried out under continuous short-day conditions, where the temperatures of dark (15 h) and light (8 h) periods were 27° and 32°, respectively. The plants of the first seven lots in Table IV were treated three times successively, namely, just before the beginning of short-day treatment, and on the first and second days under the short-day condition. The plants of the following three lots were treated with the morphactin in the same manner as the previous seven lots, but treated with indoleacetic acid or gibberellin A_3 or kinetin only on the third and tenth days of the short-day condition. The first column shows the percentage of plants in which the terminal bud flowered first. Without any chemical treatment, *Pharbitis* seedlings which received sufficient photoperiodic induction, flowered in the following manner: the terminal bud flowered first, then the flowering of lateral buds took place from the lowest to the highest.

Indoleacetic acid, gibberellin A_3 and kinetin did not alter the flowering order, while the morphactin delayed the flowering of the terminal bud and induced the lower lateral buds to flower before

Table IV

Effects of four plant growth regulators on the order of flowering in Pharbitis nil

	% Plants in which terminal bud flowered first	Days to opening of 1st flower	Plant height, cm
Control	100	26	28
IAA 100 mg/l	100	27	25
GA 100 mg/l	100	25	49
Kinetin 50 mg/l	100	25	33
M 20 mg/l	20	24	15
M 20 mg/l+IAA 100 mg/l	80	24	11
M 20 mg/l+GA 100 mg/l	67	24	36
M 20 mg/l+K 50 mg/l	40	24	16
M 20 mg/l → IAA 100 mg/l	60	26	12
M 20 mg/l → GA 100 mg/l	100	25	21
M 20 mg/l → K 50 mg/l	0	26	15

the terminal bud. This phenomenon seems to be due to one of the typical actions of morphactin reported earlier,[11] in which it retards terminal growth and stimulates the growth of lateral buds. This effect of morphactin could be reversed to a certain degree by the simultaneous application of either indoleacetic acid or gibberellin A_3. Gibberellin treatment at a later stage was more efficient than its simultaneous application with morphactin and it completely reversed the effect of morphactin. Kinetin was found to be almost inactive in reversing the morphactin action. Gibberellin A_3 also overcame the inhibitory effect of the morphactin on stem elongation as reported by many workers with different plant species.

Conclusion

Some interesting interactions between the morphactin and gibberellin A_3 on the flowering of *Pharbitis nil* have been demonstrated, although little has been understood about the mechanism of these interactions. The results suggest that more interactions of different types between morphactins and growth substances possibly exist in various phases of plant growth.

Acknowledgments

The author is indebted to Mrs. J. Chereul for her technical assistance. The morphactin was generously supplied by E. Merck Ag., Germany through Dr. Nitsch.

References

[1] Nakayama, S., *Sci. Rep. Tohoku Univ. 4th Ser.*, 1958, **24,** 137
[2] Ikeda, K., *Bull. Fac. Agric. Mie Univ.*, 1965, **31,** 1
[3] Imamura, S., Muramatsu, M., Kitajo, S. I., & Takimoto, A., *Bot. Mag., Tokyo,* 1966, **79,** 714
[4] Ogawa, Y., *Bot. Mag., Tokyo,* 1962, **75,** 92
[5] Ogawa, Y., & Imamura, S., *Proc. Jap. Acad.*, 1958, **34,** 629
[6] Ogawa, Y., *Pl. Cell Physiol.*, 1961, **2,** 343
[7] Zeevaart, J. A. D., 'Physiology of flowering in *Pharbitis nil*', 1967, p. 112 (Ed. S. Imamura) (Tokyo: Jap. Soc. Plant Physiologists)
[8] El-Antably, H. M. M., & Wareing, P. F., *Nature, Lond.*, 1966, **210,** 328
[9] Schneider, G., *Naturwissenschaften*, 1964, **51,** 416
[10] Tognoni, F., De Hertogh, A. A., & Wittwer, S. W., *Pl. Cell Physiol.*, **8,** 231
[11] Mann, J. D., Hield, H., Yung, K., & Johnson, D., *Pl. Physiol., Lancaster,* 1966, **41,** 1751
[12] Krelle, E., & Libbert, E., *Planta,* 1967, **76,** 179
[13] Khan, A. A., *Physiologia Pl.*, 1967, **20,** 306
[14] Schott, H., & Schrandolf, H., *Z. Pflphysiol.*, 1967, **56,** 387
[15] Sankhla, D., & Sankhla, N., *Biologia Pl.*, 1968, **10,** 37
[16] Harada, H., *Naturwissenschaften*, 1967, **54,** 95
[17] Sankhla, N., & Sankhla, D., *Planta,* 1967, **76,** 47

DISCUSSION

Dr. D. E. Briggs.—Recently I too have observed an unexpected interaction between gibberellic acid and morphactins in germinating barley. Increasing concentrations of morphactin progressively reduced the length of barley coleoptiles. Gibberellic acid alone increases coleoptile length, as expected. However gibberellic acid and morphactin added together inhibit coleoptile growth slightly more than morphactin added alone. Doses of morphactin that reduced coleoptile growth had no effect on α-amylase levels in the grain, and so apparently were without effect on the production and utilisation of endogenous gibberellins. (D. E. Briggs, Phytochemistry (1968), in the press).

Dr. Harada.—Thank you very much for your interesting comment. Using 4 mm. sections of *Avena* first internodes we observed that the morphactin IT 3456 could partially reverse the promoting effect of gibberellin A_3 on elongation of the sections, while the morphactin itself was slightly stimulative on the same system at low concentrations. The mode of action of morphactin must be very complex.

G. Schneider.—I would like to add some general remarks on the novel group of plant growth-regulators called 'morphactins' (*Naturwissenschaften*,, 1964, **51,** 416).

First, it must be noted that morphactins are synthetic compounds and not natural ones. If we compare the effects of the compounds with those of known natural growth-regulators like the auxins, gibberellins, cytokinins and abscisic acid, quite different behaviour is observed, depending on the specific stage of development being investigated and also on the different plant species tested. With regard to cell elongation, callus and/or adventitious root initiation, morphactins behave somewhat like auxins, while their influence on apical dominance and branching is never that of 'anti-auxins'.

As far as the extension of stem internodes is concerned, there can be observed a relatively weak gibberellin-like promotion with a few plants only, but with most other plants the opposite is observed and an 'anti-gibberellin'–like inhibition holds true; the latter applies similarly to the bolting of rosette plants (but not to flowering). Cytokinin-like activities are to be observed with leaf bud initiation and chlorophyll degradation in excised leaves (the latter as shown by the earlier work of Dr. Harada), while cell division seems to be temporarily inhibited in a somewhat 'anti-cytokinin'-like way. Concerning seed germination morphactins often seem to behave similarly to abscisic acid because of a transient retardation of the germination process without damaging the embryo; a similar picture arises with bud dormancy.

Therefore, morphactins seem to exert 'polyvalent' actions on the growth and development of plants.

Moreover, there are now some indications of interactions with others known growth-regulators, one new example of which was just shown in Dr. Harada's lecture on flowering in the short-day plant, *Pharbitis nil*.

Finally, I would like to point out two marked activity features of the morphactin regulators: (a) the compounds exert their growth regulating activity over a very wide range of concentrations, with remarkable compatibility to plants even at high dosage levels, thus having a favourable therapeutic index for diverse growth regulating purposes; (b) generally, the substances have only a short life-time inside plants as well as in soils, and they are of very low acute mammalian toxicity, thus causing no trouble from the viewpoint of toxic residues.

These features of activity look stimulating and encouraging enough to warrant further testing of morphactins within the scope of 'chemical management' of growth and development in field crops, fruit trees and ornamental plants.

MODE OF ACTION OF ABSCISIC ACID

By J. van OVERBEEK

(*Institute of Life Science, Texas A&M University, U.S.A.*)

Abscisic acid is a naturally occurring hormone which inhibits plant growth. At the Shell Agricultural Laboratory at Modesto, California the mode of action of this hormone using duckweed (*Lemna minor*) has been determined. Concentrations of the order of 10^{-6}M inhibit the growth of this plant completely and reversibly. The cytokinin benzyladenine, but not auxin nor gibberellin, counteract the inhibitory effect of the dormin. Thus it has been possible to turn growth of *Lemna* off and on by manipulating the abscisic acid and benzyladenine (in parts per billion quantities) in the medium. By analysing the synthesis of nucleic acids on MAK columns it was found that abscisic acid inhibits synthesis of all fractions, while benzyladenine promotes synthesis of all fractions.

Introduction

THE synthesis of abscisic acid by Cornforth and his co-workers and the availability of the material for physiological experimentation was a turning point in the understanding of growth and development of plants. Heretofore most plant physiologists held the view that it was incorrect to consider growth as the product of a balance between promotive and inhibitory hormones. This view was supported by the fact that auxin promoted growth in the lower concentration range, and inhibited growth at higher concentrations.

Opinions have now changed. For example, it has been demonstrated that growth of the floating duckweed *Lemna minor* can be controlled at will.[1] Abscisic acid (ABA, **I**) at a concentration of 1 ppm stops the growth; benzyladenine (BA, **II**) at 0·1 ppm causes the plant to grow again. Joseph Loeffler then demonstrated that the drastic reduction in growth induced by abscisic acid, is preceded by a striking reduction in synthesis of nucleic acids. He likewise found that the resumption of growth following application of benzyladenine, is preceded by a striking resumption in the rate of nucleic acid synthesis.[2] All the various kinds of nucleic acids as separated out by the MAK (methylated albumin on kieselguhr) column were affected.

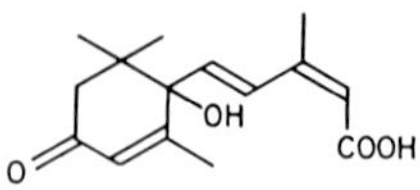

Abscisic acid
(ABA)
(**I**)

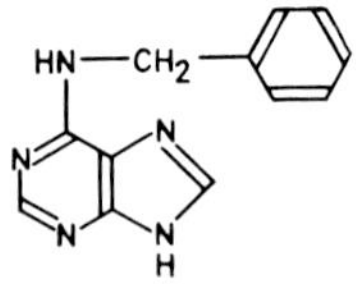

Benzyladenine
(BA)
(**II**)

There was evidence however that the second peak from the column, presumably mostly DNA, was the first to show the changes in rate of synthesis. This led to the speculation that perhaps abscisic acid slows down the rate of synthesis of DNA, that benzyladenine speeds the rate of DNA synthesis, and that these are primary effects. The change in rate of synthesis of the other nucleic acids would follow. It was also necessary to verify by further tests that the first effect of administering abscisic acid or benzyladenine was to modify the DNA synthesis.

'Stop-and-go' growth

Lemna minor was grown in sterile cultures in culture tubes in 25 ml Hoagland solution under constant fluorescent light (25 kilo lux) and constant temperature (21°±1°). An initial inoculant of three duckweed plantlets, weighing 10 mg in total, grew as shown in curve AD of Fig. 1. After one week it had grown to 100 mg. The effects of introducing ABA and BA at various stages of normal growth are shown in the other curves of Fig. 1. The introduction of 1 or 2 ppm of abscisic acid into the medium reduces the growth rate drastically. This is shown by the change in the slope of the curves. AF represents the growth when abscisic acid was added to the culture tubes at the beginning of a test. BI and CK represent growth after injection of abscisic acid to cultures that had been growing normally for a week.

The greatly depressed growth rate resulting from abscisic acid can be changed back to normal in two ways. It can be accomplished by removal of abscisic acid. This is shown in curve EG. Or, it can be accomplished by injection of benzyladenine into the dormant cultures. This is shown by the curves HL and JM.

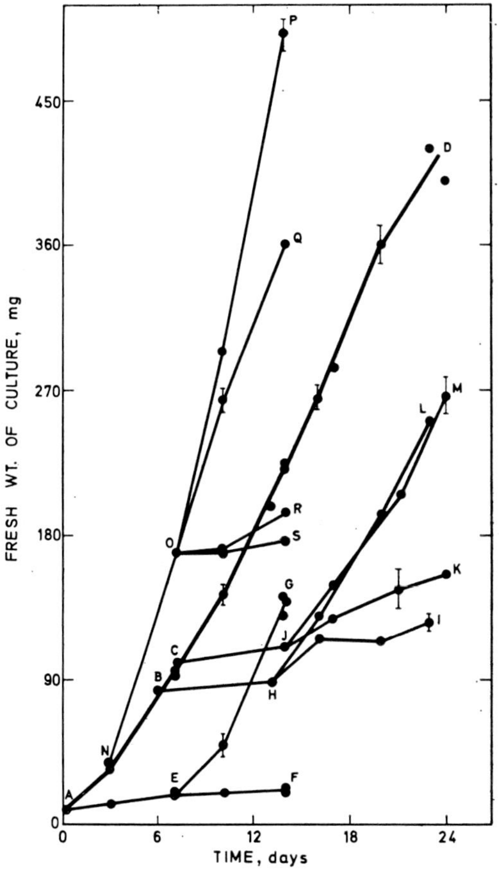

FIG. 1. *Growth of* Lemna minor *treated with benzyladenine* (*BA*) *and abscisic acid* (*ABA*)

AD = normal medium
AF = 2 ppm ABA added at commencement of test
EG = ABA removed from culture
BI = 1 ppm ABA added at point B
CK = 1 ppm ABA added at point C
HL = 0·1 ppm BA injected into dormant culture at point H
JL = 0·1 ppm BA injected into dormant culture at point J
ANP= 0·1 ppm BA added at point A
OQ = BA removed from culture at point O
OS = BA removed and 2 ppm ABA added at point O
OR = BA retained and 2 ppm ABA added at point O

'Stop-and-go' growth can also be demonstrated by first accelerating the growth rate beyond its normal limit by the application of 0·1 ppm benzyladenine. This is the optimum concentration of this cytokinin. The curve ANP results. The growth rate of such cultures then can be drastically reduced by injection of 2 ppm abscisic acid into the tubes. This occurs both where the benzyladenine was removed, as shown by curve OS, and where the benzyladenine was retained as shown by curve OR.

The graphs show clearly that 'stop-and-go' growth was achieved. It is considered this is the first time that growth of a green plant was so completely controlled by the administration of tiny quantities of plant hormones. Because both abscisic acid and cytokinins (of which benzyladenine is a synthetic representative) occur widely in nature, demonstration of 'stop-and-go' growth would, therefore, appear to reflect the way growth is controlled in nature. Any intermediate rate then might reflect a balance between inhibitory and promotive hormones.

Nucleic acid synthesis

This work was carried out by Loeffler and was published earlier.[2] Nucleic acids were extracted with phenol and detergent and separated on MAK columns. The usual fractions were found as indicated by the optical density peaks (Fig. 2). The major three peaks from left to right are presumed to be soluble RNA, DNA (largely, by analogy with peanut nucleic acids), and ribosomal RNA. Synthesis of these fractions was measured by pulse labelling with radioactive phosphate injected into the tubes. Under the methods employed changes in nucleic acid synthesis following treatment with the hormones were apparent before changes in fresh weights of the cultures. Therefore it was concluded that the changes in rates of nucleic acid synthesis were the cause of the changes in growth rates that followed. The depression of the rates of synthesis of nucleic acid fractions by abscisic acid and the enhancement of the rates of synthesis of nucleic acids by benzyladenine is evident from the curves shown in Fig. 3.

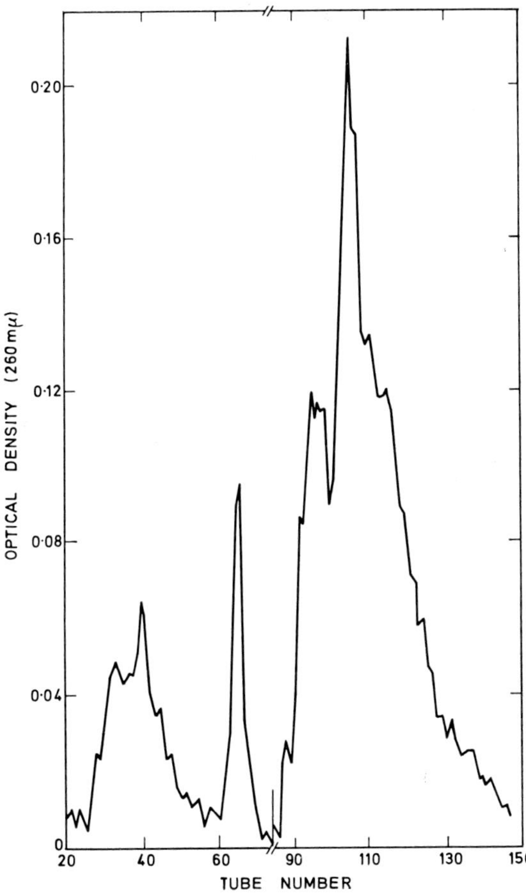

FIG. 2. *Nucleic acid fractions of normal* Lemna minor *plants following separation on a MAK column*

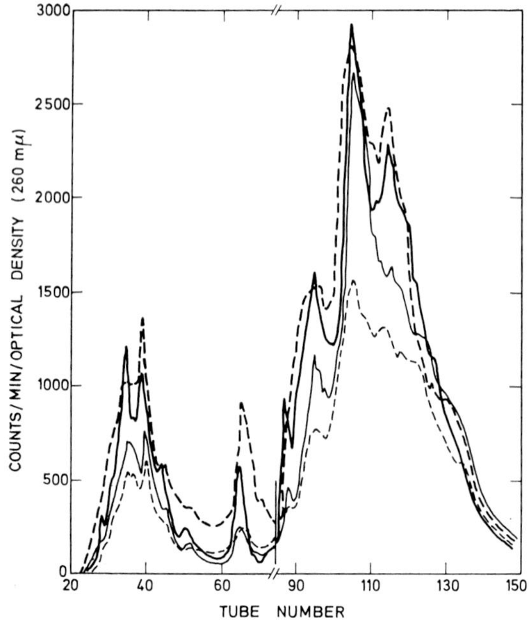

FIG. 3. *Synthesis of nucleic acid fractions as affected by ABA and BA*
——— nutrient alone
——— 1 day after addition of 1 ppm of ABA to the nutrient
– – – 3 days after addition of 1 ppm of ABA to the nutrient
— — — culture kept for 7 days on 1 ppm of ABA to which then was added 0·1 ppm of BA; plants were harvested 3 days later.

Mode of action of hormones

After assessing data[3] it was concluded that the interaction of abscisic acid and benzyladenine could not be one of competitive inhibition, where both hormones would compete for a common site of action. Instead it was found that Monod's concept[4] of allosteric affectors of an enzyme complex could account for the observations. A diagrammatic illustration showing the application to plant hormones is given in Fig. 4. The near absolute control of growth by ABA and BA suggests that the enzyme system involved must be in a very crucial metabolic step; for instance, one that would catalyse the synthesis of nucleic acids, particularly DNA. If hormones control the activity of such a sensitive step as DNA formation, it is understandable why so little hormone can affect such large masses of plant material.

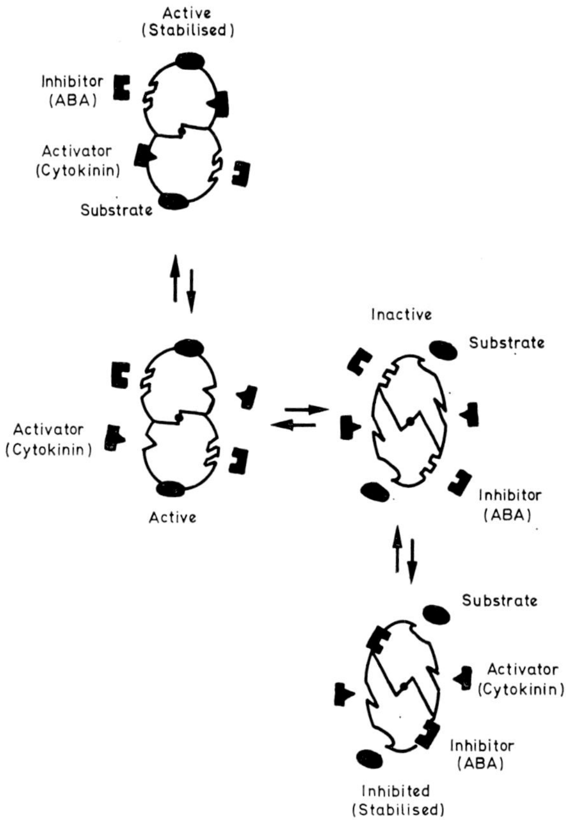

FIG. 4. *Monod's scheme of allosteric enzyme inhibition and activation adapted to the interaction of ABA and BA*

The enzyme could be a DNA polymerase, but it is not necessary to assume that the hormones would work always on one and the same enzyme system under all conditions and on all plants.

References

[1] Overbeek, J. van, & Iona M. Mason, R., in preparation.
[2] Overbeek, J. van, Loeffler, J. E. & Iona M. Mason, R. *Science, N.Y.*, 1967, **156,** 1497
[3] Overbeek, J. van, Loeffler, J. E. & M. Iona Mason, R. Proc. 6th Int. Conf. Plant Growth Substances, Ottawa, 1968. The Runge Press
[4] Monod, J., *Science, N.Y.*, 1966, **154,** 475

FOURTH SESSION

Chairman: H. HAYHURST, ESQ.
(Chairman of the Pesticides Group)

PRELIMINARY STUDIES ON THE MODE OF ACTION OF ABSCISIC ACID

By P. F. WAREING, J. GOOD, H. POTTER and A. PEARSON

(*Department of Botany, University College of Wales, Aberystwyth*)

The possible interaction between abscisic acid (ABA) and the growth promoting hormones, indole-acetic acid (IAA), gibberellic acid (GA_3) and kinetin was studied in various growth tests. No evidence of interaction between ABA and any of the other three hormones was obtained in these tests, with the following exceptions. Gibberellic acid was found to overcome the inhibitory effect of ABA on the growth of leaf sections of 'tall' (normal) maize, but not of 'dwarf' mutants; ABA appears to inhibit gibberellin biosynthesis in maize seedlings. In the lettuce seed germination test, the inhibitory effect of ABA was overcome by kinetin, but not by GA_3. The effects of ABA on RNA metabolism of radish leaf discs was investigated. Pre-treatment of leaf discs with ABA caused a marked reduction in all fractions of RNA, and this effect was particularly rapid with the poly-ribosome fractions. It is suggested that ABA may act primarily on RNA metabolism. The effect of ABA in reducing gibberellin levels in maize seedlings may be due to its inhibitory effect on the synthesis of the enzymes involved in gibberellin biosynthesis.

Introduction

THE studies which led to the isolation and identification of an abscission-accelerating substance from young cotton fruits, which was called abscisin-II,[1,2] are now well known. Parallel studies on a growth inhibitor ('dormin') extracted from the leaves of *Acer pseudoplatanus* showed that it was identical with abscisin-II,[3] the proposed structure of which was confirmed by synthesis.[4] It was also shown that the abscission-accelerating substance from yellow lupins[5] and the growth inhibitor from plum shoots[6] is identical with abscisin-II. Subsequently it has been shown that this substance is present in a wide range of plant tissues.[7] It has now been agreed that abscisin-II should be known as *abscisic acid* (ABA), but this name should not, of course, be taken as implying any specific physiological function in the plant.

ABA is highly active as a growth inhibitor in various tests,[8] and it is active in several other systems.[9] Moreover, it appears to have opposite effects to those of the three main types of growth-promoting hormones, auxins, gibberellins and cytokinins, in several tests, as will be shown below. These facts, together with its widespread occurrence, suggest that ABA may function as a natural plant growth regulator. There is evidence that ABA may act in the regulation of bud and seed dormancy[7,10,11] and in certain other developmental processes. It is of interest, therefore, to attempt to determine the possible mode of action of ABA.

The physiological antagonism between ABA and β-indolylacetic acid (IAA) is shown in certain tests; for example, in the acceleration of abscission of petioles of cotton seedlings[12] and in the *Avena* coleoptile section test.[13,14] On the other hand, in the induction and breaking of dormancy, ABA and GA_3 have opposing effects[9,10] and natural dormancy regulation may involve an interaction between these two substances. ABA also antagonises the effect of GA_3 in stimulating the synthesis of amylase in barley endosperm.[8,15] Antagonism between ABA and the synthetic cytokinins, kinetin and benzyl adenine, is shown in their effects on senescence of radish leaf discs[16] and of fronds of *Lemna minor*,[17] and in the germination of lettuce seed.[16]

Now it is clear that the simple observation that two growth substances have opposite effects in a given physiological system does not necessarily indicate that they are acting at the same point or even on the same pathway. Clearly, in so complex a process as growth, quite different partial processes may be affected by the two hormones. In this latter case, an examination of the responses to the two substances in combination at various concentrations should indicate that they are acting independently. It is necessary, therefore, to examine first whether there is any evidence of an interaction (in the statistical sense) between the two hormones, or whether their action is independent. If there is evidence of interaction, then it becomes necessary to study the nature of this interaction, of which several possible types can be envisaged.[18] The first part of this paper reports the results of experiments carried out to determine whether or not ABA interacts with the three growth-promoting substances, IAA, GA_3 and kinetin, in various tests.

The second part of the paper deals with experiments on the effect of ABA on RNA metabolism in leaf discs of radish.

Interaction between abscisic acid and growth-promoting hormones in various tests

Methods

The interaction between ABA and the growth-promoting hormones was studied in the following standard growth tests:

(i) *Avena* coleoptile test.
(ii) Dwarf, maize leaf section test.[19]
(iii) Lettuce hypocotyl test.[20]
(iv) Pea epicotyl test.[21]
(v) Lettuce seed germination test.

The effects on leaf senescence were carried out using discs cut from mature leaves of radish (*Raphanus sativus*) or dandelion (*Taraxacum officinale*) and placed in petri dishes containing the test solutions. The dishes were maintained at 20° in the dark for five days, at the end of which time the leaf discs were extracted with 80% ethanol and the chlorophyll content of the extract determined by measuring the optical density at 665 mμ.

The solutions of the various hormones were made up in aqueous solution. For these experiments synthetic ($\pm$)-abscisic acid was used, kindly supplied by Shell Research Ltd.

Results

Interaction between abscisic acid and indole-acetic acid

A series of experiments was carried out to study the possible interaction between ABA and IAA in the *Avena* coleoptile test. ABA alone is highly active as a growth inhibitor in this test (Fig. 1). The shape of this response curve is not affected by the addition of 1 ppm IAA at all ABA concentrations (Fig. 1). When various ABA and IAA concentrations were used in combination, it again appeared that the two growth substances were acting quite independently and analysis of variance gave no evidence of any interaction.

Interaction between abscisic acid and gibberellic acid

ABA has been found to antagonise the effect of GA_3 in various bioassay tests for gibberellin activity, including the lettuce hypocotyl, the dwarf maize leaf section, the oat leaf section and the pea epicotyl tests. In most of these tests, ABA is not very active as a growth inhibitor when used alone, probably because the endogenous growth without added gibberellin is itself low, and hence any growth inhibitory effect of ABA is not revealed. When

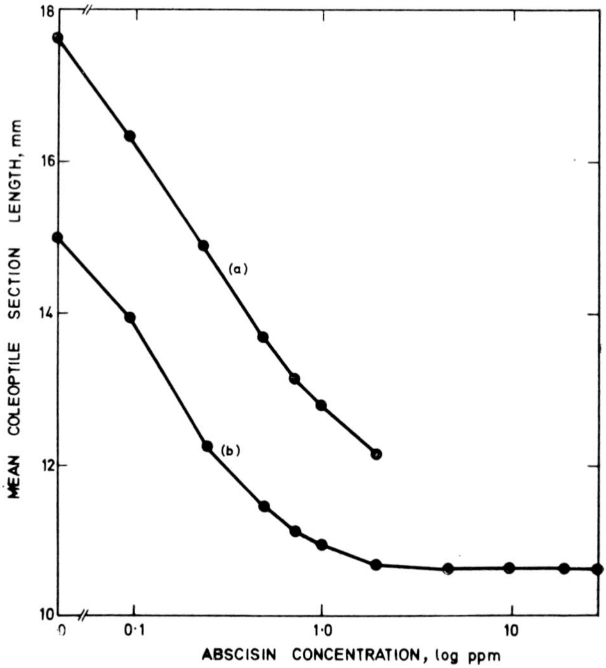

FIG. 1. *Response of oat coleoptile sections to various concentrations of ABA, in the presence and absence of IAA*
(a) with IAA at 1 ppm (b) without IAA

used in combination with added GA_3, ABA does have some growth inhibitory activity (Figs. 2–4), but it is relatively low and statistical analysis revealed no evidence of any significant interaction with GA_3 in any of these tests.

The oat coleoptile test is sensitive to added gibberellins under certain conditions, but the response is somewhat variable from occasion to occasion. In earlier tests with partially purified ABA from *Acer pseudoplatanus* leaves, it was reported[8] that the inhibitory effect of ABA in the oat coleoptile test could largely be overcome by GA_3, suggesting that these two substances may be interacting competitively. Since pure, synthetic (±)-ABA has become available, these experiments have been repeated, but it was not found possible to confirm the earlier results and no statistically significant interaction between ABA and GA_3 could be obtained (Fig. 5). The tests were also repeated with partially

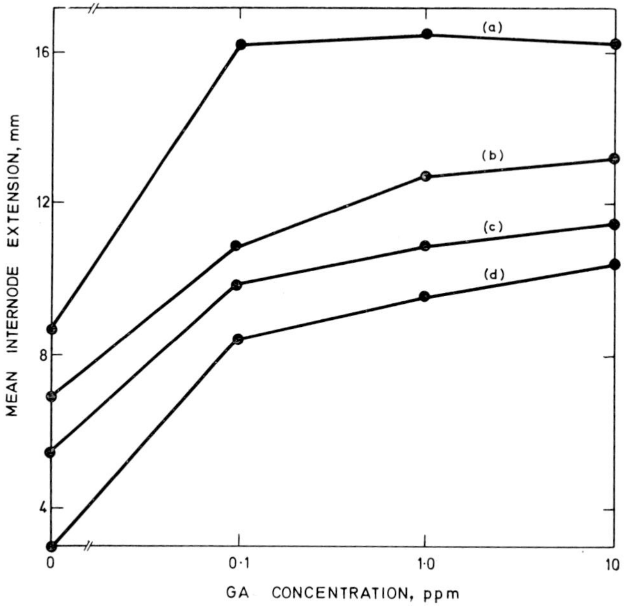

FIG. 2. *Effects of various concentrations of ABA and GA_3, alone and in combination, on growth of epicotyl sections of pea (c.v. 'Meteor')*
ABA concentration, ppm: (a) Nil (b) 1·75 (c) 3·50 (d) 7·50

purified endogenous (+)-ABA, and although the inhibition by ABA and GA_3 could be partially overcome by GA_3, the effect was not so great as in the earlier experiments and there was no significant interaction. It would seem therefore, that ABA and GA_3 act independently in the coleoptile test

Although no evidence of an interaction between ABA and GA_3 could be obtained using leaf sections of 'dwarf-1' mutant maize, it was found possible to overcome the effects of ABA by GA_3 when 'tall' (normal) maize segregants were used (Fig 6). The significance of these results is discussed below (p. 204).

Interaction between abscisic acid and kinetin

Kinetin has been shown to stimulate the growth of very young coleoptiles.[22] It has also been found that some responses to kinetin can be obtained with older sections, and the interaction between ABA and kinetin, in the presence of constant levels of IAA and

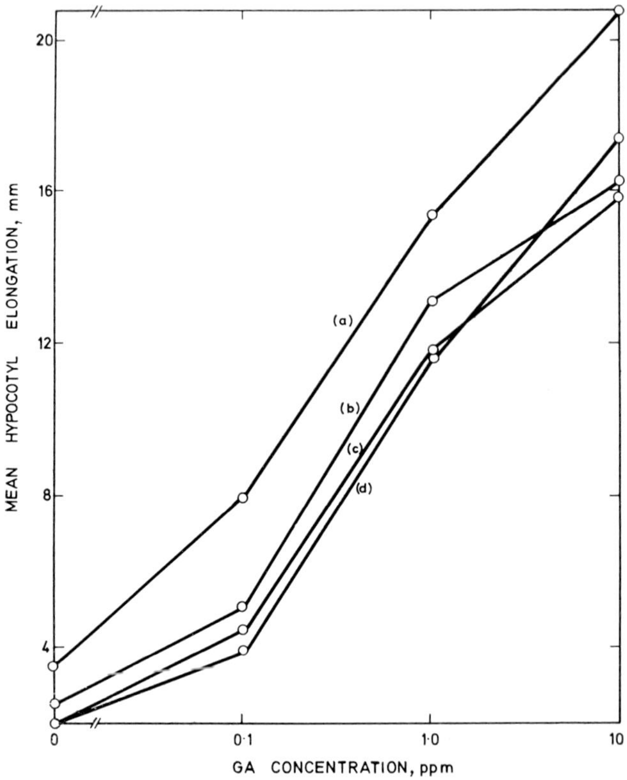

FIG. 3. *Effects of various concentrations of ABA and GA_3, alone and in combination, on growth of hypocotyls of lettuce seedlings (c.v. 'Arctic King')*

ABA concentration, ppm: (a) Nil (b) 0·25 (c) 0·50 (d) 0·75

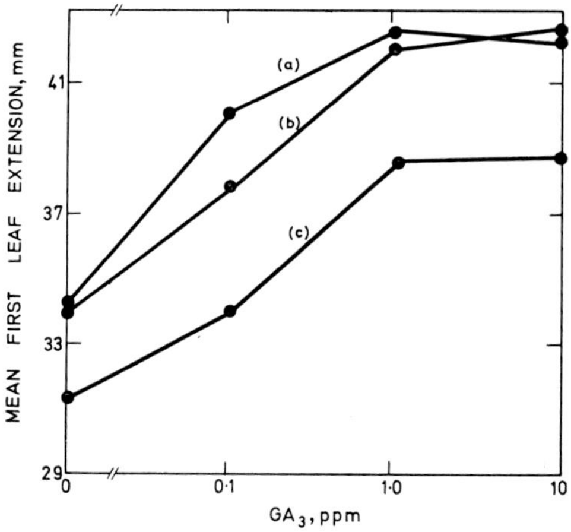

FIG. 4. *Effects of various concentrations of ABA and GA_3, alone and in combination, on growth of leaf sections of 'dwarf 1' mutant maize*
ABA concentration, ppm: (a) Nil (b) 1·0 (c) 10·0

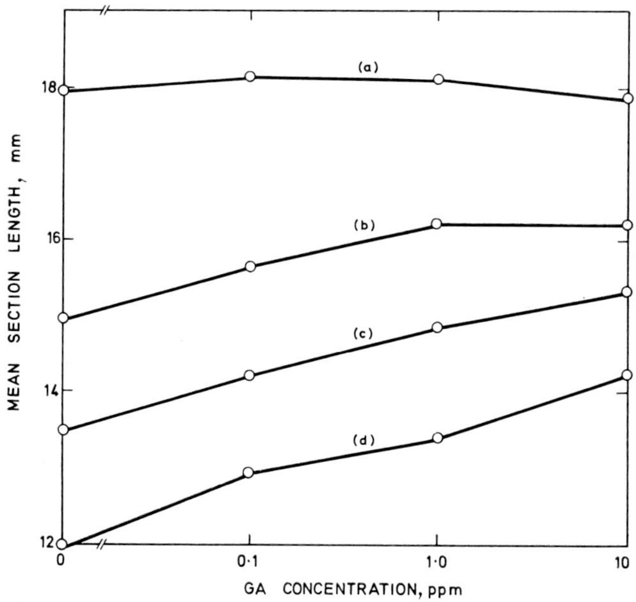

FIG. 5. *Effects of various concentrations of ABA and GA_3, alone and in combination, on the growth of oat (cv. 'Brighton') coleoptile sections*
ABA concentration, ppm: (a) Nil (b) 0·25 (c) 0·75 (d) 1·50

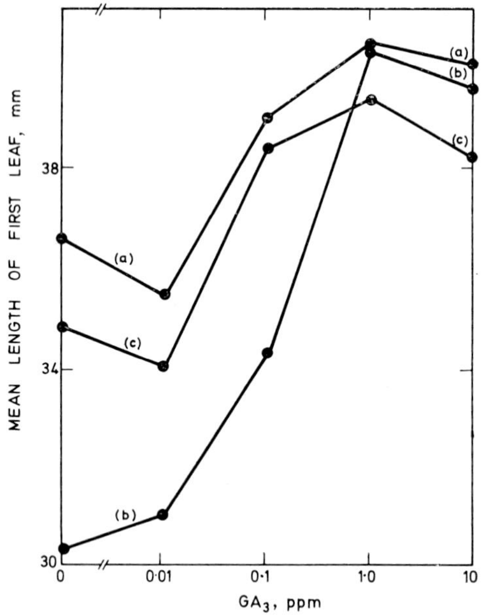

FIG. 6. *Effects of various concentrations of ABA and GA_3, alone and in combination, on growth of leaf sections of normal ('tall') maize*
ABA concentration, ppm: (a) Nil (b) 0·5 (c) 5·0

GA_3, has been studied. Again there was no evidence of any interaction between ABA and kinetin (Fig. 7).

It has been found that ABA accelerates the senescence of leaf discs.[9,16] Kinetin is known to delay leaf senescence in many species, while GA_3 is effective in certain other species. The possible interaction between ABA and kinetin was studied, using leaf discs of radish, but no evidence of any specific interaction between the effects of the two hormones was obtained. On the other hand, Aspinall *et al.*[16] found that the effects of low concentrations of ABA on the senescence of radish leaf discs were overcome by kinetin, but not the effects of higher concentrations.

On the other hand, ABA inhibits the germination of seeds of lettuce c.v. 'Great Lakes' (a non-light requiring variety) and it was possible to overcome this effect by added kinetin (Fig. 8), but not by GA_3. Similar results have been reported by Sankhla & Sankhla recently.[23]

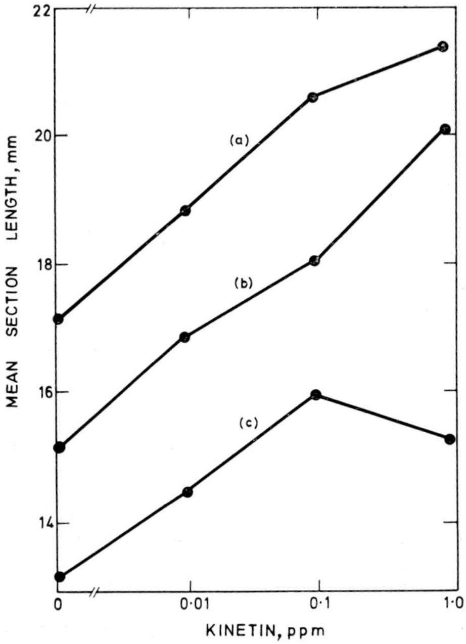

FIG. 7. *Effects of various concentrations of ABA and kinetin, alone and in combination, on growth of oat leaf sections (c.v. 'Brighton')*
ABA concentration, ppm: (a) Nil (c) 0·25 (b) 2·50

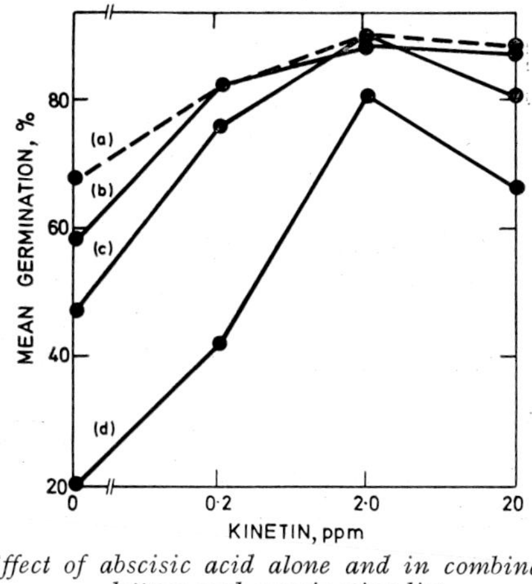

FIG. 8. *Effect of abscisic acid alone and in combination in the lettuce seed germination list*
ABA concentration, ppm: (a) Nil, (b) 0·2, (c) 2·0, (d) 20

Effects of abscisic acid on RNA metabolism

ABA has been shown to inhibit the synthesis of amylase and protease[15] in barley aleurone, i.e. it inhibits protein synthesis. However, ABA appears not to act as a general inhibitor of protein synthesis in barley aleurone, whereas it does appear to act in this way in discs of *Xanthium*[24] and oat mesocotyl sections (Poulson, R., & Saunders, P. F., unpublished results) since it reduces the incorporation of ^{14}C-leucine in these tissues. These observations suggest that ABA may inhibit some step in RNA metabolism. It was therefore decided to investigate the effect of ABA on RNA metabolism in leaf discs of radish, since it stimulates senescence in this material.

Materials and methods

Fully mature, green leaves of radish were used. The leaves were first sterilised by treating for 90 seconds in 50% 'Chloros' (sodium hypochlorite), and then rinsing in water. Discs were cut from the leaves under sterile conditions and these were placed in sterile petri dishes containing either distilled water or ABA solution (20 ppm). After a pre-treatment period varying from $5\frac{1}{2}$ to 16 h, the discs were transferred to a solution containing 200–300μ Ci of ^{32}P-orthophosphate, and the discs were incubated for a further period of 3 h, after which they were extracted for RNA.

Extraction and fractionation of RNA

A modification of the phenol partition method of Kirby[25] was used for extraction of RNA. Approximately 1 mg of RNA was obtained as a precipitate, and this was dissolved in 1·5 ml of Tris/HCl buffer. Fractionation of the RNA was carried out by centrifuging at 100,000 g at 5° for 4 h on a sucrose gradient.

Radioactivity was determined in an automatic scintillation spectrometer, using Bray's liquid scintillator.

Extraction and separation of ribosomal fractions

In the studies on the effect of ABA on the ribosomal fractions of radish leaves, 50 leaf discs (13 mm diameter) were pretreated by floating on water or ABA solution (10 ppm) and maintained in the light for 15 h.

The method of ribosomal extraction was a modification of that used by Chroboczek & Cherry[26] and Wettstein *et al.*[27] The leaf tissue was ground by pestle and mortar at −16° using 3 volumes of buffer A. The material was filtered through two layers of cheesecloth and then centrifuged at 20,000 g for 15 minutes at 0°. The pellet was discarded and the supernatant was layered over a discontinuous sucrose gradient (0·5, 1·6, 2·0M) containing buffer B.

Buffer A.	Buffer B.
Sucrose 0·25M	Sucrose 0·5, 1·6, 2·0M
Tris 0·05M, pH 7·5	Tris 0·05M
$MgCl_2$ 0·01M	$MgCl_2$ 0·01M
KCl 0·015M	KCl 0·015M
β-mercaptoethanol 0·005M	β-mercaptoethanol 0·005M
Deoxycholate 0·5%	

The gradient was centrifuged at 105,000 g for 3 h at 0°. The supernatant was discarded and the pellet, containing the ribosomes, was resuspended in 1 ml of buffer B. The equivalent of 6 optical density units (260–290 mμ in 1 cm light path) were layered on a 5 ml sucrose density gradient (10–34%), containing 0·01M $MgCl_2$, 0·015M KCl and 0·05M Tris, pH 7·5. The gradient was centrifuged in a M.S.E. Superspeed 50 swing out head at 30,000 rpm for $1\frac{1}{4}$ h at 0°. The gradient was fractionated with an Isco Model 180 Density Gradient Fractionator and 220 U-V Analyser in conjunction with a M.S.E. 10 ml tube piercer.

Results

In a number of experiments, not only with radish leaves but also with *Taraxacum*, *Nasturtium* and potato leaf discs, it has been found that pretreatment with ABA for varying periods up to 20 h results in a significant reduction in total RNA and in phenol-extractable RNA (Table I). Fractionation of this RNA by gradient centrifugation, by chromatography on MAK columns and by electrophoresis on acrylamide gel has indicated that this reduction in response to ABA extends to all fractions, and particularly to ribosomal RNA.

Table I

Effect of abscisic acid on levels of phenol-extractable RNA in radish leaf discs

Time	Phenol-extractable RNA*, mg	
	Water treatment	ABA treatment**
4 h	0·38	0·30
20 h	0·37	0·22

*RNA Determined by optical density
** 60 radish leaf discs (13 mm diameter) were treated with 10 ppm ABA for 20 h

Experiments on the incorporation of ^{32}P-orthophosphate into RNA have indicated that pretreatment with ABA also markedly reduces the incorporation. Fractionation of the RNA on a sucrose gradient showed that treatment of radish leaf discs with ABA had a marked inhibitory effect on the incorporation of ^{32}P into the various fractions. Pretreatment of the leaf discs for $5\frac{1}{2}$, 10 and 16 h respectively gave increasing inhibition of incorporation of ^{32}P into all fractions, but the effect of $5\frac{1}{2}$ h treatment was very much less with the soluble than with the ribosomal fractions (Fig. 9).

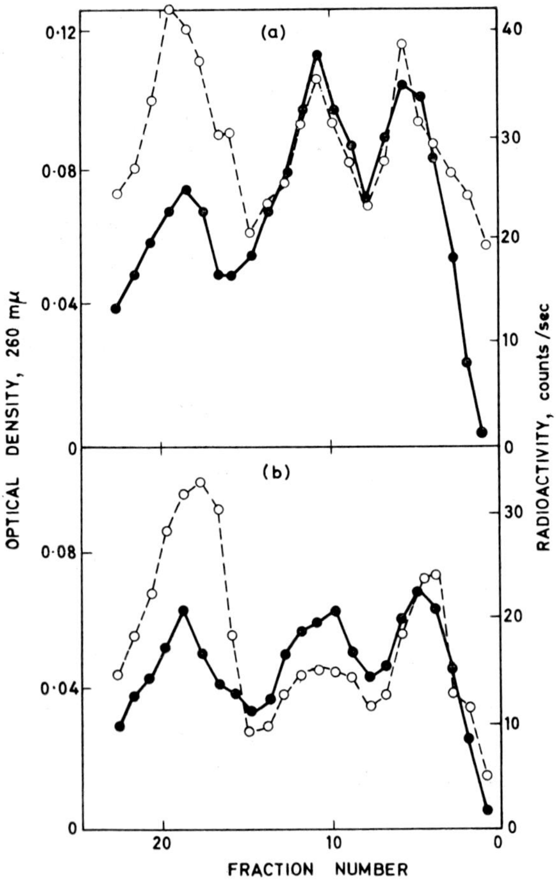

FIG. 9. *Sucrose density gradient analysis of RNA extracted from sterile radish leaf discs, pretreated for $5\frac{1}{2}$ h*
(a) Control, pretreated with water (b) Pretreated with ABA, 20 ppm
●——● Optical density ○ — — — ○ Radioactivity

Thus, there were indications that one of the earliest effects of ABA is on the incorporation into the ribosomal RNA.

In further experiments the ribosomal fractions were extracted and separated by the technique described above. The effect of pretreatment with ABA for 15 h is shown in Fig. 10. In addition to the main peak representing monoribosomes, some four other subsidiary peaks may be recognised in the fractions from the 'water' controls, which apparently represent polyribosome fractions. It was found that if the ribosome material was pretreated with RNA-ase for 5 minutes before layering on the final sucrose density gradient, the monoribosome peak was slightly reduced and the other peaks were markedly reduced, suggesting that the latter probably represent polyribosomes. It is seen from Fig. 10 that pretreatment with 10 ppm ABA for 15 h leads to a reduction in the polyribosome peaks and also to a reduction in the monoribosome fraction.

These various results provide good evidence that ABA has marked effects on the RNA metabolism of leaf discs, but whether these are direct or indirect effects requires further investigation.

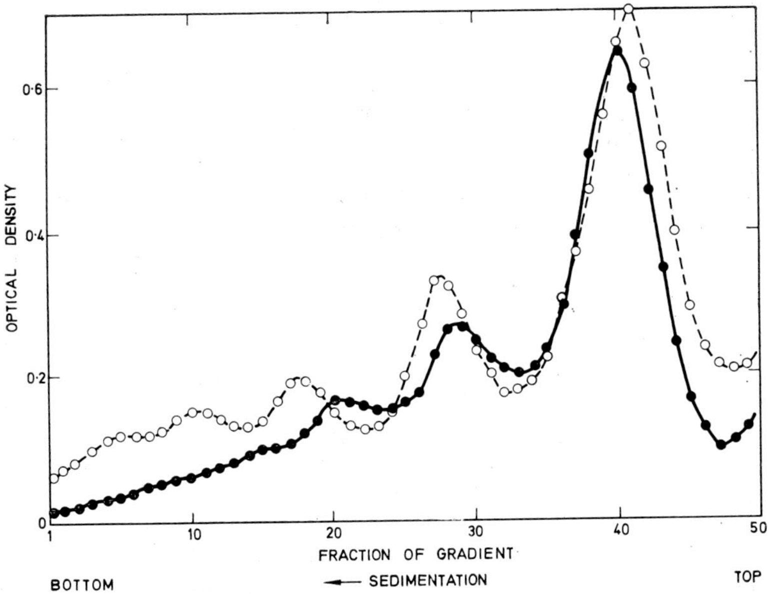

FIG. 10. *Sucrose density gradient analysis of ribosomal RNA from sterile radish leaf discs, pretreated for 15 h*

○ — — — ○ Control, pretreated with water

● —— ● Pretreated with ABA, 10 ppm

Discussion and conclusions

Extensive studies on the possible interaction between ABA and IAA, GA_3 or kinetin in various growth tests seemed to indicate that these hormones were acting independently, except in the two cases discussed below. Thus, it would seem that ABA does not show direct competitive interaction with any of these three hormones in the growth tests used. These conclusions agree with those of Milborrow,[28] and of Rothwell & Wain,[14] who used the yellow lupin factor (now known to be ABA), and concluded that there is a non-competitive interaction between ABA and IAA in the wheat coleoptile test.

In contrast with the results in most of the tests is the finding that the effects of ABA on lettuce seed germination are effectively overcome by kinetin, but not by GA_3. These results agree with those of Sankhla & Sankhla,[23] but Aspinall *et al.*[16] were able to overcome the effects of low concentrations of ABA by both kinetin and GA_3. Several authors[16,17,29] have reported that the effects of low concentrations of ABA can be overcome by GA_3 and/or kinetin but that if the concentration of ABA is too high, no reversal occurs. It is difficult to interpret this type of interaction, which does not conform with any simple pattern found in enzyme kinetics.

The fact that the effects of ABA on the growth of maize leaf sections can be largely overcome by GA_3 with the normal type, but not with the dwarf-1 mutants can be interpreted by postulating that ABA does not interact directly with GA_3, but that it affects gibberellin biosynthesis, for which evidence has been obtained.[18] Thus, if it is assumed that gibberellin biosynthesis in dwarf-1 mutants is genetically blocked at some point, then application of ABA cannot affect the endogenous levels of GA_3, which is apparently absent because of the genetical block. Hence, ABA and GA_3 should show independent effects with dwarf maize. However, in the 'normal' plants, ABA will result in reduced endogenous levels of GA_3, and part of the inhibitory effect of ABA will be due to this effect, which should be overcome by exogenous GA_3. It must be assumed, however, that the 'residual' inhibitory effect of ABA on growth of both normal and dwarf maize leaf sections, which is not overcome by exogenous GA_3, is due to its effect on some process other than gibberellin biosynthesis. Thus, it would appear that ABA does not interact competitively with GA_3, but that in maize it may inhibit the biosynthesis of GA_3. It remains uncertain whether this effect of ABA on gibberellin biosynthesis is of general occurrence.

The fact that the effects of ABA can be overcome by GA_3 in certain systems, including bud dormancy, and by kinetin in other systems, such as lettuce seed germination, suggests that ABA has rather specific effects, but that these differ from one plant system to another. Moreover, the fact that a certain part of the inhibitory effect of ABA cannot be overcome by IAA, GA_3 or kinetin in, for example, the oat coleoptile test, indicates that ABA affects certain growth processes which are apparently not under direct hormone control.

Turning now to a consideration of the effects of ABA on RNA metabolism, Chrispeels & Varner[15,29] found that in the barley aleurone layer ABA inhibits the synthesis of amylase and yet does not appear to inhibit overall RNA synthesis in this system. Thus, ABA appears to have a rather specific effect on the synthesis of certain enzymes in the barley aleurone layer.

By contrast, it has been shown above that application of ABA to radish leaf discs leads to a rapid reduction in the incorporation of ^{32}P into RNA and that fractionation of the RNA by density gradient centrifugation indicates that this reduction is shown by all fractions but seems to occur first in the ribosomal RNA. Extraction and fractionation of the ribosomal fractions by centrifugation gave a clear indication that treatment of radish leaf tissue with ABA results in a rapid reduction in the levels of polyribosomes and in monoribosomes. Thus, ABA appears to have a striking effect on RNA metabolism in radish leaf discs, but clearly it is not possible from the results obtained so far, to draw any conclusions regarding the precise point of action of ABA on RNA metabolism.

From their studies on the effect of ABA on nucleic acid metabolism in cultures of *Lemna minor*, van Overbeek *et al.*[17] concluded that ABA acts primarily on the synthesis of DNA and that it affects RNA synthesis secondarily. However, in radish leaf discs DNA synthesis is not detectable, as might be expected in mature tissue, and the effect of ABA on RNA synthesis may be more direct.

References

1 Okhuma, K., Lyon, J. L., Addicott, F. T., & Smith, O. E., *Science, N.Y.*, 1963, **142,** 1592

2 Okhuma, K., Addicott, F. T., Smith, O. E., & Thiessen, W. E., *Tetrahedron Lett.*, 1965, **29,** 2529

3 Cornforth, J. W., Milborrow, B. V., Ryback, G., & Wareing, P. F., *Nature, Lond.*, 1965, **205,** 1269

4 Cornforth, J. W., Milborrow, B. V., & Ryback, G., *Nature, Lond.*, 1965, **206,** 715

[5] Cornforth, J. W., Milborrow, B. V., Ryback, G., Rothwell, K., & Wain, R. L., *Nature, Lond.*, 1966, **211,** 742
[6] Barlow, H. W. B., Hancock, C. R., & Lacy, H. J., *Proc. Fourth Int. Conf. Plant Growth Regulation*, 1961, p. 127 (Ames: Iowa University Press)
[7] Milborrow, B. V., *Planta*, 1967, **76,** 93
[8] Thomas, T. H., Wareing, P. F., & Robinson, P. M., *Nature, Lond.*, 1965, **205,** 1270
[9] El-Antably, H. M. M., Wareing, P. F., & Hillman, J., *Planta*, 1967, **73,** 74
[10] Eagles, C. F., & Wareing, P. F., *Physiologia Pl.*, 1964, **17,** 697
[11] Jackson, G. A. D., & Blundell, J. B., *J. hort. Sci.*, 1963, **38,** 310
[12] Addicott, F. T., Carns, H. R., Lyon, J. L., Smith, O. E., & McMeans, J. L., *Proc. Fifth Int. Conf. Plant Growth Regulation*, 1964, p. 687 (Paris: Edition du C.N.R.S.)
[13] Robinson, P. M., & Wareing, P. F., *Physiologia Plant.*, 1964, **17,** 314
[14] Rothwell, K., & Wain, R. L., *Proc. Fifth Int. Conf. Plant. Growth Regulation*, 1964, p. 363 (Paris: Edition du C.N.R.S.)
[15] Chrispeels, M. J., & Varner, J., *Nature, Lond.*, 1966, **212,** 1066
[16] Aspinall, D., Paleg, L. G., & Addicott, F. T., *Aust. J. Biol. Sci.*, 1967, **20,** 869
[17] Overbeek, van J., Loeffler, J. E., & Mason, M. I. R., *Science, N.Y.*, 1967, **156,** 1497
[18] Wareing, P. F., Good, J., & Manuel, J., *Proc. Sixth Int. Conf. Plant Growth Regulation*, 1968 (in press)
[19] Frankland, B., & Wareing, P. F., *Nature, Lond.*, 1962, **194,** 313
[20] Frankland, B., & Wareing, P. F., *Nature, Lond.*, 1960, **185,** 255
[21] Phillips, I. D. J., & Jones, R. L., *Planta*, 1963, **63,** 269
[22] Wright, S. T. C., *Nature, Lond.*, 1961, **190,** 699
[23] Sankhla, N., & Sankhla, D., *Physiologia Plant.*, 1968, **21,** 190
[24] Osborne, D. J., *Symp. Soc. Exp. Biol.*, 1967, p. 305 (Cambridge: University Press)
[25] Kirby, J., *Biochem. Biophys. Acta*, 1962, **55,** 545
[26] Chroboczek, H., & Cherry, J., *Biochem. Biophys. Res. Comm.*, 1965, **20,** 474
[27] Wettstein, F. O., Staehelin, T., & Noll, H., *Nature, Lond.*, 1963, **197,** 430
[28] Milborrow, B. V., *Planta*, 1966, **70,** 155
[29] Chrispeels, M. J., & Varner, J., *Pl. Physiol., Lancaster*, 1967, **42,** 1008

DISCUSSION

Mr. H. W. B. Barlow.—The speed with which abscisic acid-induced inhibition can be relieved by simple washing out with water—at least in coleoptile sections—makes it difficult to see how an effect on nucleic acid metabolism can be involved in at least this inhibition.

Professor Wareing.—I agree that, at first sight, it is difficult to imagine that a response involving RNA metabolism would occur rapidly, but we really have no idea of the rates of RNA synthesis in higher plants. Certainly in bacteria, for example, large changes in RNA can occur in 5–10 minutes.

Dr. D. J. Osborne.—Do you think we should still refer to abscisic acid (ABA) as an abscission-accelerating substance? The experiments we reported at Ottawa showed quite clearly that ABA at low concentrations has no effect on abscission or excised abscission zones, and at higher concentrations it inhibited abscission. This we correlated with a progressive inhibition of protein synthesis with increasing concentration of ABA. If however, ABA is supplied after a long period, during which time the induction (or activation) of enzymes responsible for separation has taken place, then ABA *does* accelerate abscission. But this can be explained by the ABA enhancement of senescence of the regions above the separation zone, and ABA enhancement of senescence certainly occurs, as you have just described.

Professor Wareing.—I think your suggestions regarding the way in which ABA may cause accelerated abscission are very convincing. Certainly they would fit in well with the fact that ABA enhances senescence rate and with its effects on RNA metabolism and enzyme synthesis. Hitherto, when ABA has been described as an abscission-accelerating substance, one had in mind the 'pharmacological' effects, rather than any specific mode of action. However, we may have to stop referring to ABA as an abscission-promoter in future. The term *abscisic acid* does not imply any specific physiological function, but reflects the history of its discovery.

STUDIES ON THE MECHANISM OF ACTION OF INDOLEACETIC ACID

By D. D. DAVIES, B. D. PATTERSON AND A. J. TREWAVAS

(School of Biological Sciences, University of East Anglia, Norwich, Norfolk)

Experiments are reported showing that indoleacetic acid (IAA) stimulates the synthesis of proteins and nucleic acids. The time sequence indicates that these effects are subsequent to elongation of the cell wall. An attempt has been made to determine if IAA has a selective or general effect on protein and nucleic acid synthesis. Evidence presented in this paper indicates that the action of IAA is selective on both protein and nucleic acid synthesis. In particular, the results reported suggest the synthesis of IAA induced messenger RNA, the stimulation of heavy ribosomal RNA synthesis and an increase in polysome production.

Introduction

TREATMENT of plant tissues with naturally occurring or synthetic auxins often results in profound physiological changes.[1] The molecular basis for some of these changes could lie in the controlled alteration of the activity of cellular enzymes. In recent years, work on bacterial systems has provided workers in the hormonal field with a number of models and testable hypotheses on the mechanism of control of enzyme activities.

In this paper an attempt is made to establish that auxin produces a change in the balance of enzymes by a 'de novo' synthesis of protein and to examine the molecular basis of this reaction in terms of current concepts in molecular biology.

Results and discussion

Sections from the etiolated third internode of Alaska pea stems were used to examine the effects of indoleacetic acid (IAA) on protein synthesis. Half of the plant material was supplied with a ^{14}C labelled amino acid and IAA in low concentration (10^{-5}M). The other half was supplied with the same amino acid, but labelled with tritium (^{3}H). At various time intervals the two tissue batches were combined, washed, homogenised and the protein extracted and fractionated by column chromatography. The $^{14}C/^{3}H$ ratios were measured in each fraction. The criterion of protein synthesis used in these experiments has been the incorporation of labelled amino acids (both lysine and leucine have been used) into material which is resistant to ribonuclease and deoxyribonuclease and which behaves like protein on Sephadex-gel and DEAE-'cellulose chromatography'. If the effect of auxin is to increase or not to change the rate of incorporation of amino acids into all proteins, the $^{14}C/^{3}H$ ratio of each of the protein fractions should be identical. If on the other hand the $^{14}C/^{3}H$ ratio in all protein fractions is not identical then it can be concluded that IAA has increased or decreased the rate of synthesis of some proteins with respect to others.

An example of this type of experiment is shown in Fig. 1. The effect of IAA on the ratio $^{14}C/^{3}H$ when leucine was used as appre-cursor is compared with that using lysine. Deoxycholate was used to extract protein from a crude nuclear and plastid fraction referred to as the $5000 \times g$ fraction. The protein was fractionated on a column of Sephadex G-150. Controls were run omitting IAA to establish the 'noise level' of the experiment. The effect of prein-

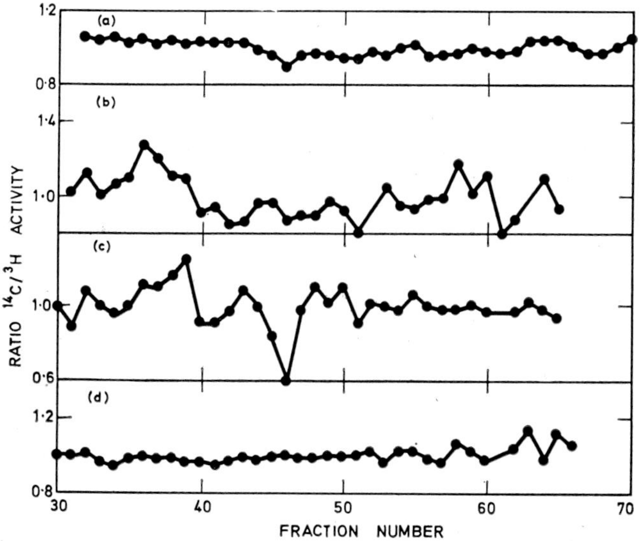

FIG. 1. *Sephadex gel G-150 separations of deoxycholate-soluble nuclear protein from tissue incubated in the presence and absence of IAA*

Batches of pea stem tissue were incubated in the presence of labelled lysine in (a), (c) and (d) and leucine in (b) for 2 h. In separation (a), two tissue batches were incubated under identical conditions except that one batch received tritiated lysine and the other ^{14}C lysine. In separations (b), (c) and (d) indole-acetic acid (2×10^{-5}M) was added to the ^{14}C amino acid at the beginning of the experiment. In separation (d), actinomycin D (10 μg/ml) was added to both batches of tissue 1 h before addition of amino acids and IAA.

cubation with actinomycin D (10 μg/ml) for 1 h was also examined.

These and similar experiments[2] establish that IAA produces alterations in $^{14}C/^{3}H$ ratio in particular protein fractions and that actinomycin D abolishes these alterations. These results could be interpreted as indicating that continued DNA-dependent RNA synthesis was necessary for auxin to produce differential synthesis of protein. Two points of caution should however be noted: (i) the lack of definite evidence concerning the specificity of action of actinomycin D in plants and (ii) the possibility that differential labelling of protein may not reflect differential synthesis of protein. If two proteins are considered in a steady state with grossly different rates of synthesis and consequently different rates of breakdown, doubling the rate of synthesis of both does not necessarily result in an equal increase in the amount of label in both proteins.[3] However, this explanation is unlikely since it requires that the rate of protein synthesis be large relative to the pool size of the protein.[4]

Accepting the evidence that IAA alters the pattern of protein synthesis leads in terms of current ideas to an examination of the effects of IAA on the synthesis of RNA. The effect of added IAA on the synthesis of RNA in etiolated pea stems was investigated using $^{32}P_i$ as the isotopic precursor. The specific activity of the RNA (counts/min/μmole nucleotide) was determined by electrophoresis of the component nucleotides after alkaline hydrolysis. The results are shown in Fig. 2 in which the separate bars represent separate experiments. The existence of a lag period about an hour before auxin stimulates RNA synthesis is apparent.

Essentially the same experiment was performed using labelled

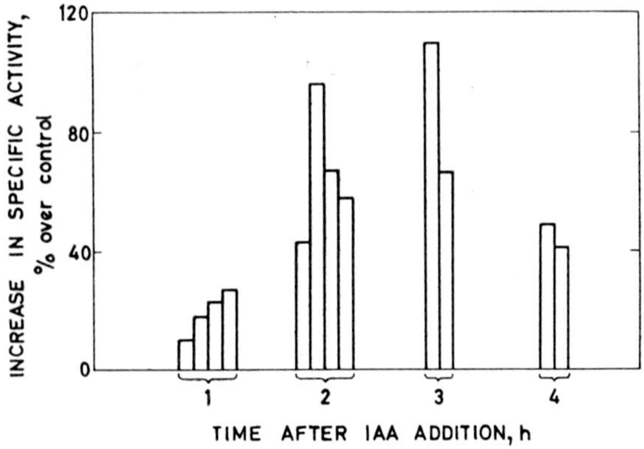

FIG. 2. *Effect of IAA on the specific activity of RNA in pea stem sections using $^{32}P_i$ as a precursor*

orotic acid as the precursor. The results are shown in Fig. 3; the more careful analysis of the first hour after IAA addition further supports the view that there is a distinct lag before IAA stimulates RNA synthesis. Separation of the nuclei from the cytoplasm[5] prior to the measurement of incorporation of label into RNA, permits this detection of which cell fraction first responds to IAA. The results presented in Fig. 4 indicate that the IAA-stimulated RNA appears in the nucleus prior to its appearance in the cytoplasm.

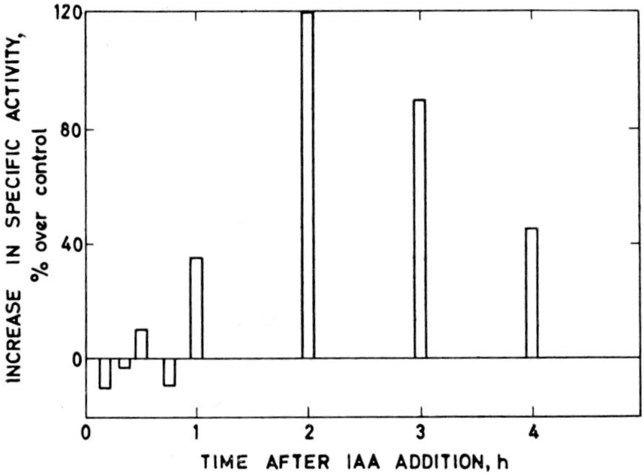

FIG. 3. *Effect of IAA on the specific activity of RNA in pea stem sections using 6-^{14}C-orotic acid as a precursor*

Specific activity was determined by alkaline hydrolysis of the RNA and separation of the component nucleotides by paper electrophoresis.

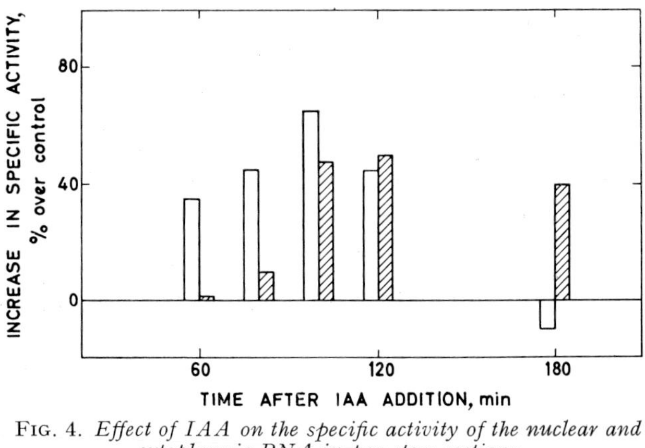

FIG. 4. *Effect of IAA on the specific activity of the nuclear and cytoplasmic RNA in pea stem sections*

□ = specific activity increase of nuclear RNA. ▨ = specific activity increase of cytoplasmic RNA.

These changes in RNA metabolism occur in the essential absence of increases in the uptake of the two isotopic precursors into the tissue. Further, the net level of RNA remains unchanged for the first few hours of incubations of pea stem sections in IAA. Thus the data shown in Figs. 3 and 4 indicate that after an initial lag period, IAA increases the rate of synthesis of RNA in the nuclei of pea epicotyls. The effect of IAA on RNA synthesis was compared with the effect of IAA on the incorporation of amino acids into protein. The results shown in Fig. 5 indicate a similar lag period for both effects and suggest a causal relationship.

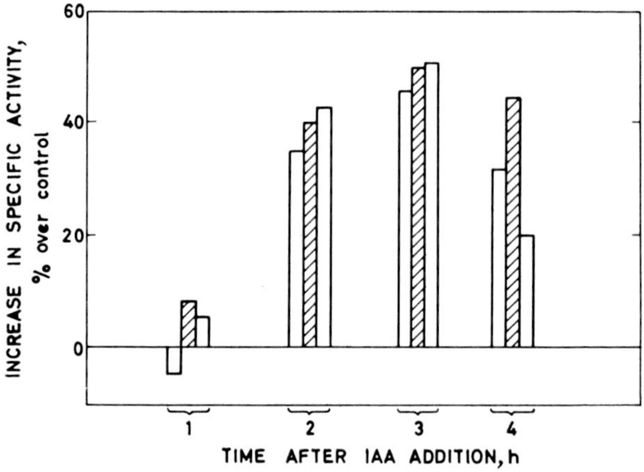

FIG. 5. *Effect of IAA on the specific activity of protein in pea stem sections*
Batches of pea stem sections were incubated in ^{14}C leucine ▨ or ^{14}C valine □ for the time periods shown above. At the end of these periods the specific activities of the protein were determined as previously described.[5]

Since the early 1960s the concept of messenger RNA and the notion of the control of messenger production have had a profound effect on biological research. The original concept,[6] operator controlling operon messenger production and the operator being controlled by a repressor, has been shown to be true for a few bacterial systems. The original theory has however been modified to allow control at the translational as well as transcriptional level. The theory was originally introduced to explain the induction and repression of enzymes in bacteria. Bacteria grow at a rate dependent, almost entirely, on the availability of external substrates. The ability to adapt rapidly to a new type of substrate or to stop manufacturing one which is present in abundance in the environment is of much greater selective advantage for bacteria than higher organisms, where control of the internal environment, tissue differentiation and maintenance of mature organs are possibly more important factors. Because of these differences the extrapolation of the concepts explaining control in bacteria to regulation of protein synthesis in higher organisms is not necessarily valid. Notwithstanding this objection the concepts developed in bacterial biochemistry have been used to attempt to explain hormonal control of protein synthesis in plants.[7]

At first an attempt was made to study the effect of IAA on messenger RNA synthesis using the technique of DNA/RNA hybridisation. This technique developed from the observation that under certain conditions denatured (i.e. single stranded) DNA and RNA will combine with each other to form a hybrid, provided the two molecules have complementary nucleotide sequences.[8] If radioactive RNA is used for the hybridisation the hybrid can be isolated and the amount of radioactivity in the hybrid is a measure of the amount of RNA hybridised. By varying the conditions of incubation it is possible to achieve a situation in which only messenger RNA is effectively being hybridised.[9]

The RNA synthesised in pea stem tissue in the presence and absence of IAA was examined by the technique of competition hybridisation. In this technique the radioactive hybrid is formed in the presence of large quantities of unlabelled RNA, from the same source. The unlabelled RNA competes with the labelled RNA in the formation of the hybrid and thus results in a lowering of the radioactive hybrid formed, producing a competition curve (Fig. 6).

This figure shows examples of competitive hybridisation of ^{32}P-labelled RNA isolated from tissue incubated in the presence and absence of IAA. In the top two curves RNA isolated from the protozoan *Paramecium aurelia* was used as a control. In the lower two curves the competitor was RNA isolated from pea stem tissue incubated in the absence of both IAA and isotope. The fact that IAA increases the amount of label appearing in the hybrid could be interpreted as showing that IAA has increased the synthesis of messenger RNA. If the synthesis of all types of messenger RNA was increased by auxin to the same extent the two lower curves should be identical in shape. To test this the percentage increase in counts hybridised due to IAA treatment is plotted in Fig. 7.

If the two curves are identical in shape this value should be a straight horizontal line. The upper curve represents the results using pea stem RNA as a competitor and the lower *Paramecium* RNA. As can be seen, when unlabelled pea RNA is used as a competitor the results do not follow the predicted pattern if IAA had merely increased the synthesis of all forms of messenger RNA.

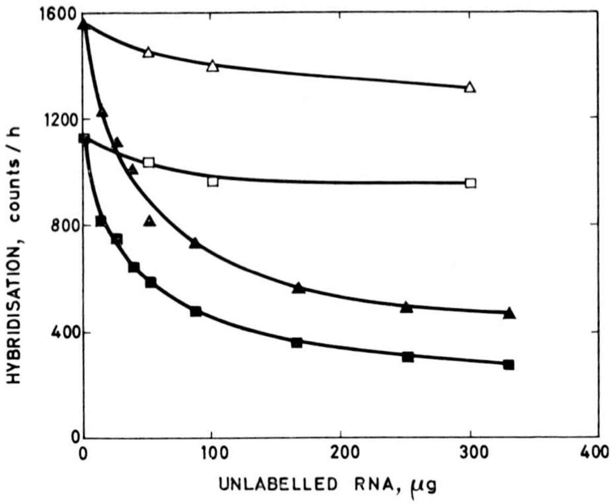

FIG. 6. *Hybridisation of pea DNA to labelled pea RNA obtained from tissue incubated in* $^{32}P_i$ *and in the presence of IAA for* 2 h

7·5 μg labelled RNA was incubated with 37 μg DNA at 42°c for 15 h in the presence of increasing unlabelled RNA levels and the hybrids collected and measured as previously described.[5] Specific activity of RNA from IAA-treated tissue = 1930 counts/h/μg RNA. Specific activity of RNA from untreated tissue = 1280 counts/h/μg RNA. All samples were counted to a standard error of less than ±2% of the count minus background.

▲ RNA from IAA treated tissue competed with unlabelled pea RNA
■ RNA from untreated tissue competed with unlabelled pea RNA
△ RNA from IAA treated tissue competed with unlabelled *Paramecium* RNA
□ RNA from untreated tissue competed with unlabelled *Paramecium* RNA

A possible interpretation of the data is that IAA stimulates the production of specific RNA 'messages' and these 'messages' are in low concentration in the untreated tissue. This IAA-stimulated messenger RNA would not be effectively competed against because of the low concentration of its counterpart in the RNA used for competition. However, as noted earlier, an apparent differential pattern of labelling could occur if there are gross differences in the rates at which individual messages approach isotopic equilibrium.[3]

The data presented in Fig. 7 shows that the increase in counts hybridised due to IAA treatment was about 35%, whereas the increase in specific activity of the total RNA was well over 50%. This suggests that IAA increases the synthesis of non-hybridising forms of RNA, i.e. ribosomal and transfer RNA, to a greater extent than it does messenger.

The incubation conditions necessary to eliminate serious inter-

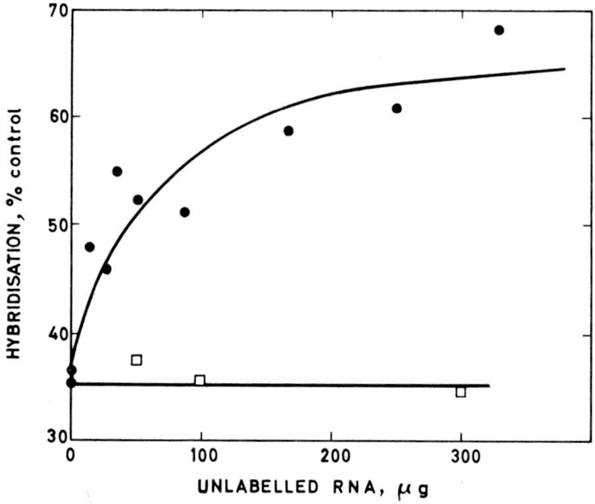

FIG. 7. *IAA-induced increase in hybridisation during competition by unlabelled RNA*

The ordinate was calculated from Fig. 6 by expressing the difference in counts hybridised between the IAA treated and untreated RNA as a percentage of the untreated RNA counts hybridised. The ordinate therefore represents percentage increase in counts hybridised due to IAA treatment of tissue.

● Auxin treated and untreated tissues competed with unlabelled pea RNA

□ Auxin treated and untreated tissues competed with unlabelled *Paramecium* RNA

ference from ribosomal RNA involves using very high RNA/DNA ratios, i.e. the proportion of the total hybrid which represents ribosomal RNA is very small. By using low ratios of labelled RNA to DNA in the incubation mixture, ribosomal RNA will form a much greater proportion of the hybrid and the effect of IAA on the labelling of ribosomal RNA can thus be studied. Fig. 8 shows the effect on counts hybridised of increasing the levels of labelled RNA in the incubation medium. The top curve represents labelled RNA from IAA-treated tissue and the lower curve labelled RNA from untreated tissue. Ribosomal RNA forms a saturating level of hybrid at 2 μg labelled RNA.[10]

These data are re-expressed in Fig. 9 as the percentage increase in counts hybridised due to IAA treatment. At very low RNA/DNA ratios, IAA increases the counts hybridised by nearly 60%. As the level of labelled RNA is increased (thus reducing the proportion of hybrid representing ribosomal RNA), the IAA-induced increase in counts hybridised markedly declines. The data is thus consistent

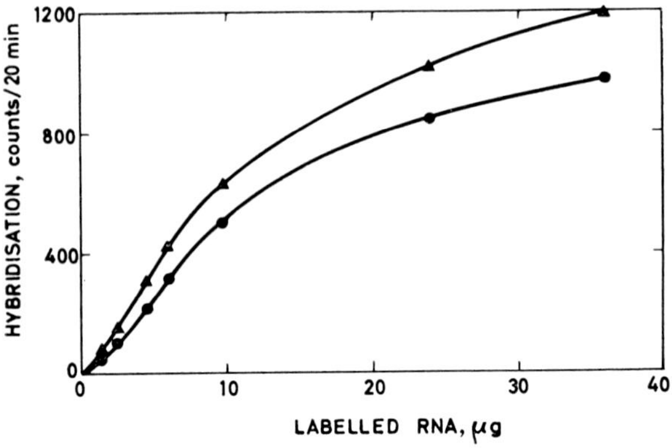

FIG. 8. *Effect of increasing levels of labelled RNA from IAA treated and untreated tissue on the level of label hybridised*

Labelled RNA was prepared using $^{32}P_i$ from pea stem tissue incubated in the presence and absence of IAA for 2 h and hybridised to 50μg denatured pea DNA.

Specific activity of RNA from IAA-treated tissue = 2260 counts/h/μg RNA. Specific activity of RNA from untreated tissue = 1380 counts/h/μg RNA. Increase in specific activity due to IAA treatment of tissue = 64%. All samples were counted to a standard error of less than ±2% of count minus background.

▲ RNA from IAA-treated tissue
● RNA from untreated tissue

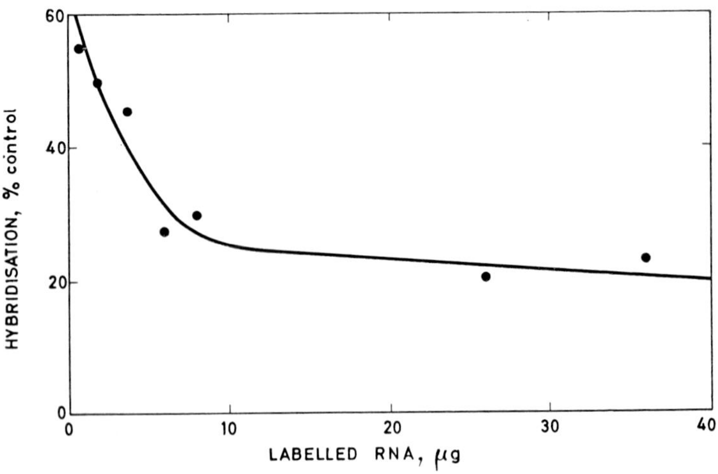

FIG. 9. *Increase in hybridisation due to IAA treatment*

The ordinate was calculated from Fig. 8 by expressing the difference in the counts hybridised between the IAA treated and untreated RNA as a percentage of the untreated RNA counts hybridised.

with the view that IAA enhances the synthesis of ribosomal RNA to a much greater extent than messenger RNA.

Further evidence for this view has been obtained by making a base analysis of the newly synthesised RNA. Epicotyls were treated with $^{32}P_i$ in the presence and absence of IAA. The RNA was extracted, hydrolysed with alkali and the component nucleotides separated. The results presented in Table I show that the ratio GMP/AMP (Guanosine Monophosphate/Adenosine Monophosphate) is increased by IAA. This change also mirrors the time course of the IAA-induced change in specific activity. Sub-cellular analysis of the tissue reveals the altered GMP/AMP ratio to be located in the mitochondrial and microsomal fractions. The sub-cellular fractions designated here have a purely operational definition and are not pure. The mitochondrial fraction will for example contain polyribosomes. It should be noted that Ingle & Key[11] were able to associate a 2,4D-induced increased in the

Table I

A. *Effect of IAA on the GMP/AMP ratio of whole cell RNA*

	GMP/AMP ratio		
	Incubation time 1 h	Incubation time 2 h	Incubation time 3 h
+IAA (GMP/AMP)	0·69±0·015*	0·91±0·04	0·87±0·034
−IAA (GMP/AMP)	0·65±0·02	0·71±0·04	0·72±0·02

B. *Effect of IAA on the specific activity and GMP/AMP ratio of RNA in subcellular fractions*

	Nuclear fraction		Mitochondrial fraction		Microsomal fraction		Supernatant fraction	
	+IAA	−IAA	+IAA	−IAA	+IAA	−IAA	+IAA	−IAA
GMP/AMP	0·85	0·81	0·71	0·60	0·89	0·74	0·61	0·60
Specific activity	7·2	4·8	2·6	1·9	3·7	2·4	7·0	3·5

The effect of IAA on the GMP/AMP ratio of the newly-synthesised RNA was determined as follows. Tissue was incubated in $^{32}P_i$ in the presence and absence of IAA for varying time periods. The labelled RNA was hydrolysed in alkali and the component nucleotides separated. Sub-cellular fractionation was carried out as previously described.[5]

* Values represent mean ± standard deviation of the mean.

GMP/AMP ratio of newly-synthesised RNA in soy-bean epicotyls with a selective increase in ribosomal RNA.

The technique of double-labelling was also used to examine the effect of IAA on the synthesis of different forms of RNA. Pea epicotyls were incubated in the presence of tritiated adenosine and ^{14}C-adenosine and IAA. After incubation the tissues were combined, the nucleic acids extracted and fractionated on a methylated-albumin kieselguhr column. The ratio of ^{14}C to ^{3}H was determined in each fraction and the results are shown in Fig. 10.

An increase in the $^{14}C/^{3}H$ ratio in any fraction indicates a selective alteration of synthesis of that fraction. Only fractions 60–140 are shown since they contain the relevant peaks. The DNA (fractions 60–70) and the two ribosomal peaks (fractions 80–115) may be clearly located from the optical density profile and the messenger (fractions 102–132) as the highly labelled back peak. The biggest alteration in the $^{14}C/^{3}H$ ratio occurs over the second

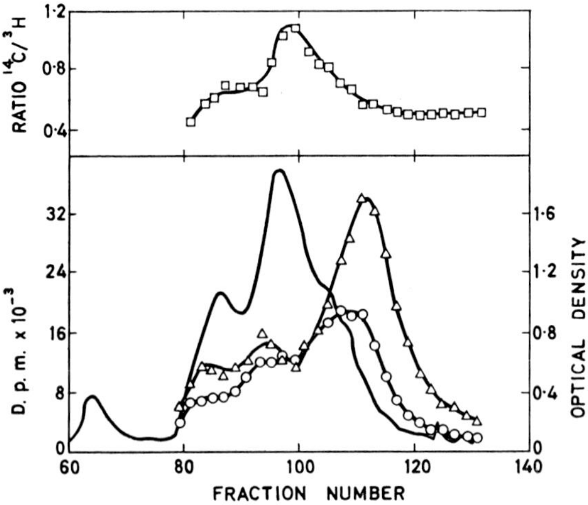

FIG. 10. *Methylated-albumin kieselguhr column chromatography of labelled RNA from stem tissue incubated in the presence and absence of IAA*

Pea stem tissue was incubated in ^{14}C-adenosine+IAA and in ^{3}H-adenosine for 2 h. The tissue was bulked and the nucleic acids extracted and fractionated. The disintegrations due to carbon and tritium were determined in each fraction. Radioactivity is only shown over fractions 80–132. No counts were detectable in the DNA region.

d.p.m. = Disintegrations per minute
□——□ Ratio $^{14}C/^{3}H$
△——△ ^{3}H
○——○ ^{14}C
———— Optical density

of the two ribosomal peaks, i.e. the heavy ribosomal RNA. However even the first (light ribosomal RNA) peak shows a higher ratio than that in the messenger region.

The effect of IAA on ribosome production has also been further examined employing the technique of density gradient centrifugation.

Fig. 11 shows sucrose density gradient profiles of ribosomes isolated from pea stem tissue incubated in the presence and absence of auxin for 12 h. The figures do not represent labelling but actual amounts as measured by their optical density. The largest peak represents the monomer and the succeeding peaks have sedimentation values which suggest they are the dimers, trimers and tetramers of the monosomes.[12] The area of the peaks under the polysomes and monosomes in this and two other experiments have been summated and the results are shown in Table II.

All the experiments indicate that auxin has increased the amount of material in the polysomes by well over 50% but the level of the monosomes remains unaltered. These data could be interpreted as

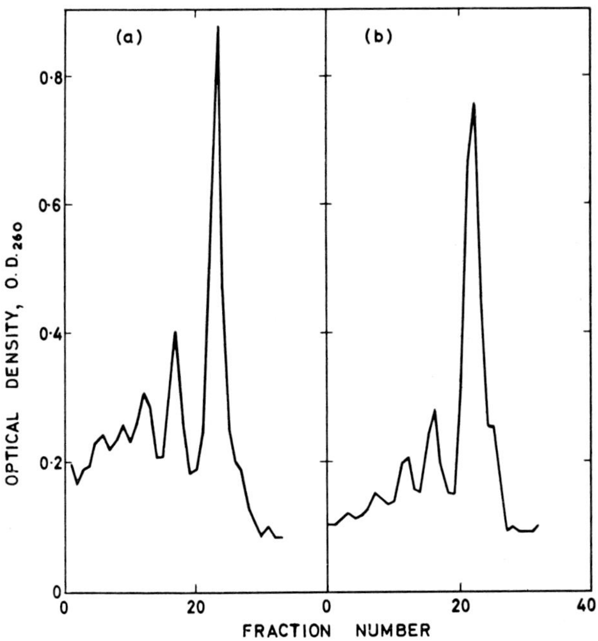

FIG. 11. *Effect of IAA on the sucrose density gradient profiles of ribosomes isolated from etiolated pea stem tissue*
Pea stem tissues incubated for 12 h (a) with IAA, (b) without IAA

Table II

Effect of auxin on the level of polysomes and monosomes in etiolated pea tissue
Values represent total $O.D._{260}$ absorption in the denoted gradient region

Experiment number	Polysomes		Monosomes	
	With IAA	Without IAA	With IAA	Without IAA
1	4·64	2·92	2·76	2·85
2	5·30	2·95	2·70	2·56
3	4·71	3·10	2·79	2·65
Mean	4·88	2·99	2·75	2·69
Increase due to IAA	63%		2·2%	

indicating that IAA has enhanced the rate of synthesis of ribosomal RNA and in turn the level of the polysomes. This explanation is complicated by the fact that incubation of tissue in sucrose also leads to a 50% increase in the level of polysomes compared with tissues incubated in water.[12] The data of Fig. 11 could also be interpreted as meaning that IAA reduces the rate of degradation of polysomes. The incorporation of labelled orotic acid into ribosomes and polysomes has therefore been examined.

Pea epicotyls were incubated with ^{14}C-orotic acid in the presence and absence of IAA for 2 h. The ribosome population was isolated, separated and the specific activity of each determined. The results are shown in Table III and suggest that IAA increases the incorporation of label into polysomes to a greater extent than into monosomes.

Table III

Effect of IAA on the incorporation of labelled orotic acid into the polysomes of etiolated pea tissue
Figures represent the specific activity in terms of disintegrations/min/μg RNA in the particular fractions denoted

	Polysomes		Monosomes	
	With IAA	Without IAA	With IAA	Without IAA
Specific activity	112	61	78	53
Increase due to IAA	86%		47%	

The significance of this observation remains to be evaluated. Possible newly synthesised ribosomes form polysomes more rapidly than 'old' ribosomes. Alternatively the increased population of polysomes is a reflection of the simultaneous production of IAA-specific messenger RNA and ribosomes or possibly by even IAA-specific ribosomes.

Conclusion

IAA has been shown to have a selective effect on the pattern of protein synthesis. Other results have suggested that this effect may be initiated in the nucleus by the synthesis of new kinds of RNA which have some of the properties of the messenger of RNA.

The increases in the relative rate of synthesis of ribosomal RNA, and the increases in polysome production, may be a primary requirement for the stimulation of protein synthesis by auxins which is observed in whole plants.[13]

Acknowledgments

The authors would like to acknowledge the gifts of samples of Actinomycin D and paramecium RNA from Dr. I. Gibson. This work was supported by A.R.C. Grant AG 83/4 to Professor D. D. Davies.

References

[1] Audus, L. J., 'Plant Growth Substances', 1959 (London: Leonard Hill)
[2] Patterson, B. D., & Trewavas, A. J., *Pl. Physiol., Lancaster*, 1967, **42,** 1081
[3] Reiner, J. M., *Arch. Biochem. Biophys.*, 1963, **46,** 80
[4] Holleman, J. M., & Key, J. L., *Pl. Physiol., Lancaster*, 1967, **42,** 29
[5] Trewavas, A. J., *Arch. Biochem. Biophys.*, 1968, **123,** 324
[6] Jacob, F., & Monod, J., *J. Molec. Biol.*, 1961, **3,** 318
[7] Van Overbeek, J., *Science, N.Y.*, 1966, **152,** 721
[8] McCarthy, B. J., *Progr. Nucleic Acid Res.*, 1965, **4,** 129
[9] McCarthy, B. J., & Hoyer, B. H., *Proc. Natl. Acad. Sci.*, 1964, **52,** 915
[10] Chipchase, M. I. H., & Birnsteil, M. L., *Proc. Natl. Acad. Sci.*, 1963, **50,** 1101
[11] Ingle, J., & Key, J. L., *Pl. Physiol., Lancaster*, 1965, **40,** 1212
[12] Trewavas, A. J., *Phytochemistry*, 1968, **7,** 673
[13] Fan, Der-Fong, & McClachlan, G. A., *Can. J. Bot.*, 1966, **44,** 1025
[14] Key, J. L., *Pl. Physiol., Lancaster*, 1964, **39,** 365
[15] Ray, P. M., & Ruesink, A. W., *Develop. Biol.*, 1962, **4,** 377
[16] Vogel, A. I., 'A Textbook of Practical Organic Chemistry', 1953 (London: Longmans)
[17] Andreae, W. A., & Ysellstein, M. W., *Pl. Physiol., Lancaster*, 1960, **35,** 225
[18] Zenk, M. H., *Colloq. Int. Centre Natl. Rech. Sci.*, 1964, **123,** 241
[19] Christiansen, G. S., & Thimann, K. V., *Arch. Biochem.*, 1950, **28,** 117
[20] Lonberg-Holm, K. K., *Nature Lond.*, 1967, **213,** 454

DISCUSSION

Dr. A. Winter.—How do the elegant experiments you have described relate to the growth of the tissue?

Dr. A. Trewavas.—We do not feel that the changes we have observed are directly related to cell extension. Very early on in this work we measured the auxin effects on fresh weight and length and it was soon apparent that no lag period in response was detectable.[5] This result would hardly have been expected if the RNA and protein changes (both with lag periods of about an hour) were direct precedents of cell extension. Incidentally, the 2,4-D–induced RNA changes in excised soy-bean hypocotyl exhibit a lag period of nearly 3 h[14] and in this system the metabolic changes appear to be more closely related to the increased cell division induced by 2,4-D rather than increased cell extension. Addition of high levels of IAA to intact stem tissue is also accompanied by large increases in RNA and protein and massive cell division[13] and we feel that the altered RNA and protein synthesis we have detected is more closely related with this cell division than with cell extension.

Dr. A. Winter.—It has been known for some 30 years that auxin can double the rate of protoplasmic streaming in *Avena* coleoptile cells in several minutes. The more recent work of Ray & Ruesink[15] has shown an irreversible extension of the cell wall in response to added IAA within 10–15 min.

Dr. A. Trewavas.—I agree that both these changes take place very rapidly. Incidentally we have tried to repeat the protoplasmic streaming response with little success.

Dr. A. Winter.—It would seem therefore that changes in nucleic acids and proteins are more closely related to the correlative effects of IAA than to extension growth. Incidentally it is possible to repeat the early experiments of Thimann & Sweeney, the conditions however are critical.

Dr. M. Hallaway.—Do you think that the breakdown of IAA to other active and inactive compounds could account in part for the lag you observe in the response to IAA, or for the fall off in response after 3 or 4 h?

Dr. A. Trewavas.—The IAA used in these experiments was recrystallised by the procedure of Vogel[16] until the product was colourless with a temperature melting range of 1°. We therefore think it unlikely that the lag period is the result of other compounds present in the indoleacetic acid. As to the second part of the question, pea tissue has been known for a number of years to have an active IAA oxidase[1] and this could account in part for the 'apparent' fall off in RNA and protein changes.

Dr. A. Winter.—Might I say that the kinetic experiments of both Andreae[17] and his co-workers and Zenk[18] working with pea tissue, indicate that conjugates of IAA are not detectable for at least 2 h after application of IAA at the concentrations of IAA you were using.

Prof. B. B. Stowe.—Would you specify the growth conditions for the peas, and the incubation medium?

Dr. A. Trewavas.—Peas were grown and stem sections incubated by the method of Christiansen & Thimann.[19]

Prof. J. van Overbeek.—How about sterility?

Dr. A. Trewavas.—Some bacteria are present in our incubation medium since we have not carried out experiments under sterile conditions. With this problem in mind we have always avoided the use of sucrose and buffers, used short incubation periods and thoroughly washed the tissue after incubation when we have been using labelled precursors. We also have some further evidence that bacterial incorporation of isotopic precursors is unlikely to play a measurable role in the RNA and protein changes we have discussed.

Lonberg-Holm[20] was able to show that contaminated seedlings exhibited extensive incorporation of label into DNA and that nearly all of this incorporation was the result of bacterial replication. When pea stem tissue is incubated under our conditions for 2 h in tritiated adenosine we have not been able to detect any label in the DNA fractions eluted from MAK columns. Incubation for 4 h results in about 1% of the total incorporated label now appearing in the DNA region and this may be the result of bacterial activity. However this proportion of the label would compare favourably with the results for sterile seedlings shown by Lonberg-Holm.[20]

Prof. J. P. Nitsch.—In relation to Dr. van Overbeek's comment concerning the possible effects of bacteria, may I mention that we have done similar experiments under sterile conditions, using tobacco pith tissue. Electron-microscopy has shown that IAA induces the synthesis of a large amount of ribosomes with numerous polysomes. Thus our results substantiate Professor Davies' conclusions.

Dr. A. Trewavas.—It should be pointed out of course in many of our experiments in which an initial low speed centrifugation has been performed, e.g. isolation of polysomes, any bacterial contamination will have been removed.

RELATION OF THE STRUCTURE OF FATTY ACID DERIVATIVES TO THEIR ACTION AS CHEMICAL PRUNING AGENTS*

By H. M. CATHEY and G. L. STEFFENS

(*Crops Research Division, U.S. Department of Agriculture, Beltsville, Maryland, U.S.A.*)

Fatty acid derivatives with chain lengths of C_8, C_{10}, and C_{11} were highly active in selectively killing the terminal buds of seven different species of plants. The axillary buds were unaffected and grew at nearly the same rate as axillary buds on plants pruned manually. The chemicals thus substituted for a mechanical operation essential in the culture of many plants. Fatty acid derivatives with chain lengths shorter than C_9 or longer than C_{11} were less effective. On an equal molar basis the propargyl compounds (propargyl-2-octyl sulphite) were more active than fatty alcohols (1-undecanol), which in turn were more active than fatty acid esters (methyl decanoate). The selective action of the fatty acid derivatives was dependent upon the use of a suitable amount and type of surfactant.

Introduction

FATTY acid derivatives have been reported as active inhibitors of seed germination[1] and bacterial growth,[2] and as simulators of auxin and gibberellin action in growth promotion of pea stem sections.[3] They were found to be killers of axillary buds (suckers) of tobacco by Tso[4] and Tso *et al.*[5] The C_8, C_9, C_{10}, and C_{12} methyl esters were most effective in killing the lateral meristems without damaging the mature leaf or stem tissue. The chemicals acted only as contact killers and were effective only when they came in direct contact with the meristematic tissues to be killed. The chemicals were neither translocated, nor were they altered by the metabolic processes of the treated plants.[6] The selective kill of meristematic tissues by fatty acid esters on a wide range of plants was first reported by Cathey *et al.*[7] These authors suggested that the fatty

*This is a report on the current status of research on growth regulating chemicals. It does not contain recommendations for their use, nor does it imply that the uses discussed have been registered. All uses of these chemicals must be registered by appropriate State and Federal agencies before they can be recommended.

Mention of a trademark name or a proprietary product does not constitute a guarantee or warranty of the product by the USDA, and does not imply its approval to the exclusion of other products that may also be suitable.

acid esters should be called 'chemical pruning agents' since they killed only the growing tip without damaging the remaining parts of the plant. Lateral meristems developed at nearly the same time as those on plants from which the stem apex was removed by hand. The chemical treatment substituted for an established process used by many growers to stimulate the development of side branches.

This paper reports on three groups of compounds which are effective and selective killers of terminal meristems without damaging the axillary meristems.

Materials and methods

Test plants

The test emulsions of chemical pruning agents were applied to *Chrysanthemum morifolium* Ramat. c.v. 'Improved Indianapolis Yellow' plants. The plants were grown from rooted cuttings in a pad and fan air-cooled greenhouse in photoperiodic conditions that assured their remaining in a vegetative condition throughout the experiments. These conditions consisted of natural photoperiods and 4 h of incandescent light of 20 ft-c from 10 p.m. to 2 a.m. daily (long photoperiods).

Similar experiments were conducted with many of the fatty acid derivatives on apple, English holly, hemlock, petunia, red maple, and weigela. These plants were grown under long photoperiods that assured their remaining in active growth throughout the experiment. The night temperature was maintained at a minimum of 17°. All plants were fertilised and watered with routine greenhouse practices. All experiments were conducted at least in duplicate; there were at least 6 treated stems per treatment, and the results reported are averages.

Spray emulsions

The chemical pruning agents are insoluble in water. They were dispersed in water with the aid of various types of surfactants. The action of the chemical pruning agents in selectively killing meristematic tissues depends on the use of a suitable amount and type of surfactant. The development of colloidal emulsions which were opalescent and stable was essential for the desired action of the chemical pruning agents. The technique for selection of the surfactant is beyond the scope of this paper and will be reported in detail elsewhere. The surfactants used were: polyoxyethylene (20) sorbitan monolaurate (Tween-20), polyoxyethylene (20) sorbitan mono-oleate (Tween-80), and polyoxyethylene–alkyl aryl ether–alkyl aryl sulphonate blend (Atlox 3403). The following families of chemical pruning agents were sprayed on plants as emulsions:

Fatty acid esters: (a) methyl ester of hexanoate (C_6), octanoate (C_8), nonanoate (C_9), decanoate (C_{10}), undecanoate (C_{11}), dodecanoate (C_{12}), myristate (C_{14}), and palmitate (C_{16}); (b) different esters of decanoate: methyl (C_1), ethyl (C_2), propyl (C_3), isopropyl (C_3) n-butyl (C_4), tert-butyl (C_4), n-pentyl (C_5), n-hexyl (C_6); (c) esters with a total of 10 carbons: n-nonyl formate (C_9–C_1), n-octyl acetate (C_8–C_2), n-hexyl butyrate (C_6–C_4), n-pentyl pentanoate (C_5–C_5), n-butyl hexanoate (C_4–C_6), n-propyl heptanoate (C_3–C_7), ethyl octanoate (C_2–C_8), and methyl nonanoate (C_1–C_9); (d) various positions of ester group on fatty acid chain: ethyl-α-ethyl hexanoate, ethyl-α-methyl octanoate, ethyl-α-ethyl octanoate, ethyl-α-propyl octanoate; ethyl-γ-ethyl hexanoate, and ethyl-α-propyl pentanoate.

Fatty alcohols: (a) various chain lengths: 1-hexanol (C_6), 1-heptanol (C_7), 1-octanol (C_8), 1-nonanol (C_9), 1-decanol (C_{10}), 1-undecanol (C_{11}), 1-dodecanol (C_{12}), 1-tetradecanol (C_{14}), 1-hexadecanol (C_{16}), and 1-octadecanol (C_{18}).

Propargyl compounds: (a) various chain lengths: propargyl-n-hexyl sulphite, paropargyl-n-octyl sulphite, propargyl-n-nonyl sulphite, propargyl-n-decyl sulphite, and propargyl-n-dodecyl sulphite; (b) structural modifications: propargyl-n-octyl sulphite, propargyl-2-octyl sulphite, propargyl-n-decyl sulphite, and propargyl decanoate.

For tests with chrysanthemum plants, the concentration of the fatty acid esters was 0·11, 0·135, and 0·16M with 3% Tween-20. The fatty alcohols were 0·08 and 0·16M with 1·5% and 3% Tween-80, respectively, and the propargyl compounds were 2, 3, 4, and 5×10^{-2}M with 1% Atlox 3403. Emulsions were prepared by weighing the test chemical and surfactant into a container and mixing thoroughly. A small volume of water equal to the original volume was added to thicken and to form a gel, and water was then added gradually to make up the emulsion to the required volume. The emulsions were used as soon as prepared.

Test method

The emulsions were applied with an ordinary throat atomiser. The entire top of the plant, including terminal meristem and surrounding leaves of various stages of maturity, were sprayed until the emulsion began to glisten on the foliage. The plants were not sprayed until run-off because excessive drip of the liquid down the stem sometimes caused girdling of the stem at the soil line.

The lateral growth was harvested 21 days after treatment and was weighed to the nearest 0·01 g on a top-loading balance. Results are expressed as a percentage of the growth from manually pruned plants. Leaf damage was extimated (x-limited to xxxx-severe damage). Lateral buds were counted on the top of the stem.

Results

The C_8, C_9, C_{10}, C_{11}, and C_{12} methyl esters of the fatty acids were highly active on vegetative chrysanthemum plants (Fig. 1).

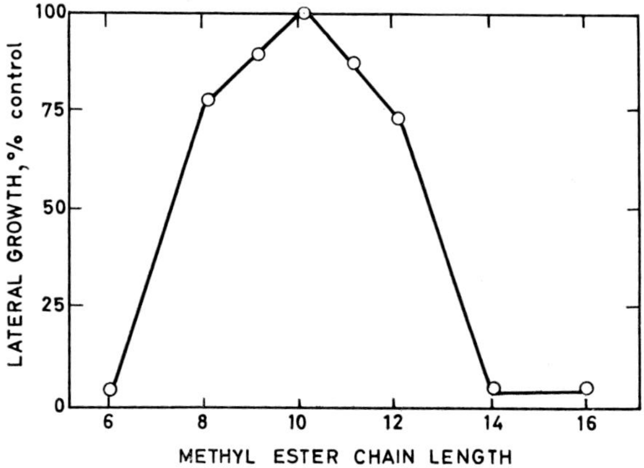

FIG. 1. *Axillary bud growth of 'Improved Indianapolis Yellow' chrysanthemum*
21 days after treatment with 0·16M fatty acid methyl ester and emulsified with 3% Tween-20
Control: manually pruned = 100%

The terminal meristem turned black after about 15 minutes, flaccid within 30 to 60 minutes, and tan within the 3 to 5 days (Fig. 2). The C_6, C_{14}, and C_{16} methyl esters were inactive at 0·16M concentrations. The most effective compound was the C_{10} methyl ester. The weight of the lateral branches was similar to the amount removed from plants with the terminal meristem removed manually but the chemically pruned plants produced an average of 1·5 more lateral branches (4·5) than did plants pruned manually (3·0). Little or no leaf damage was observed on plants treated with compounds containing less than 11 carbons. The C_{12} and C_{14} methyl esters caused a glazing of the leaves; the damage persisted throughout the duration of the experiment. The C_{16} methyl ester left a visible white deposit on the leaves. Partially expanded leaves at the time of treatment developed a slight marginal burn, the leaves exhibited abnormal ridging of the veins and the blades were puckered in appearance (Fig. 3).

FIG. 2. *Close-up of terminal meristem of chrysanthemum killed from spray application of a fatty acid derivative*
Photographed 1 day after treatment

FIG. 3. *Axillary bud growth of plants of 'Improved Indianapolis Yellow' chrysanthemum treated with methyl esters of varying chain lengths*
Sprayed with emulsions containing 0·16M of the methyl ester of from left to right, C_6, C_8, C_{10}, C_{12}, C_{14}, and C_{16} fatty acids
Photographed 21 days after treatment

The methyl (C_1), ethyl (C_2), and n-propyl (C_3) esters of decanoic acid were equally effective on chrysanthemum plants. Isopropyl decanoate was much less active than the n-propyl decanoate (Fig. 4). Isopropyl decanoate applied at a concentration of 0·16M did not actually kill the terminal meristems, although their development was delayed. Apical dominance was lost for several days, so that lateral branches began to develop. The terminal meristem then resumed growth. The n-butyl (C_4), n-pentyl (C_5) and n-hexyl (C_6) esters also caused a temporary halt in the growth of terminal meristems. Lateral branches developed partially and stems thus had terminal and several lateral branches developing simultaneously.

Compounds containing 10 carbons with ester linkages between various carbons were tested on chrysanthemum plants. When applied at a concentration of 0·11M, only n-nonyl formate (C_9–C_1) and methyl nonanoate (C_1–C_9) were active. When the concentration was increased to 0·135M, n-octyl acetate (C_8–C_2) was also active. A further increase in concentration to 0·16M made ethyl octanoate (C_2–C_8) active (Fig. 5). The fatty acid esters containing less than 8 carbons in a chain were relatively inactive.

The position and the number of carbons making up the ester

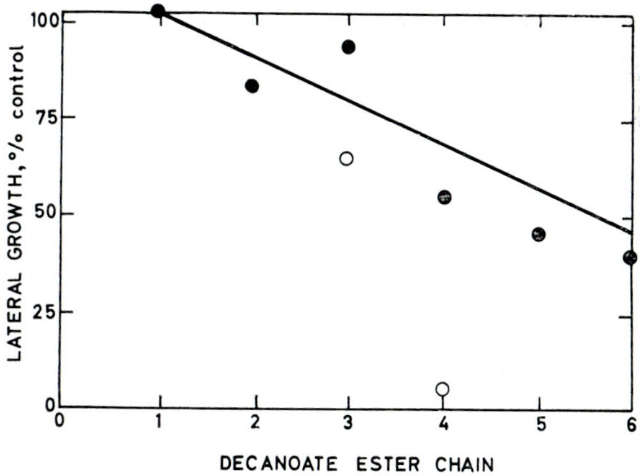

FIG. 4. *Axillary bud growth of 'Improved Indianapolis Yellow' chrysanthemum treated with decanoate esters*

21 days after treatment with 0·16M of the fatty acid decanoate esterfied with various ester chain lengths and emulsified with 3% Tween-20

Control: manually pruned = 100%

● Straight chain ○ Branched chain

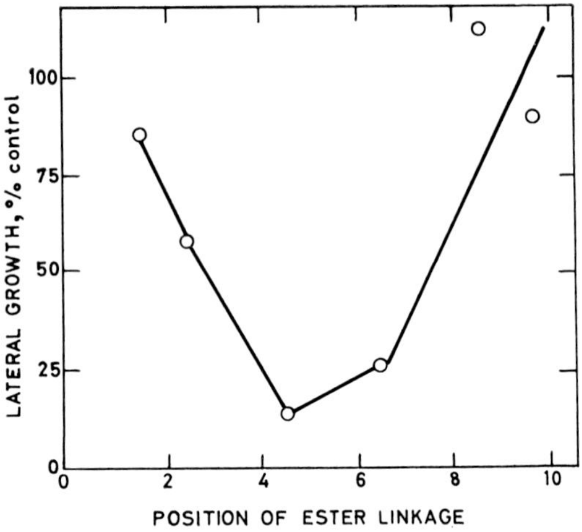

FIG. 5. *Axillary bud growth of 'Improved Indianapolis Yellow' chrysanthemum treated with fatty acid esters of varying ester linkages*
21 days after treatment with 0·16M of fatty acid esters with the esterification at the position indicated and emulsified with 3% Tween-20
Control: manually pruned=100%

altered its effectiveness. Ethyl-α-ethyl hexanoate (10C), ethyl-γ-ethyl hexanoate (10C), ethyl-α-propyl pentanoate (10C), ethyl-α-methyl octanoate (11C), ethyl-α-ethyl octanoate (12C), and ethyl-α-propyl octanoate (13C) were inactive when tested on chrysanthemum plants at 0·16M.

The C_8, C_9, C_{10}, and C_{12} fatty alcohols were highly active as chemical pruning agents on chrysanthemum plants at 0·08M (Fig. 6). The C_{11} alcohol, 1-undecanol, killed both the terminal meristem and also many of the axillary buds but the stem and mature leaves were unaffected. Eventually the lateral shoots developed from basal nodes of the stem. The C_{12} and C_{14} fatty alcohols caused a glazing of the leaves; the C_{14} treated leaves were shiney, with desiccated interveinal areas, marginal burn and cupping of leaves which were partially expanded at treatment time. The C_{16} alcohol emulsion left a white residue on the foliage without damaging the plant. The C_6 and C_7 alcohols were inactive at a concentration of 0·08M.

A second set of chrysanthemum plants was treated with 0·16M emulsions of the various fatty alcohols. The C_9, C_{10}, and C_{11} fatty alcohols destroyed the terminal meristem and in addition, the axillary buds were killed if the emulsions came in contact with them, mature leaves and stem were unaffected. The C_8 and C_{12}

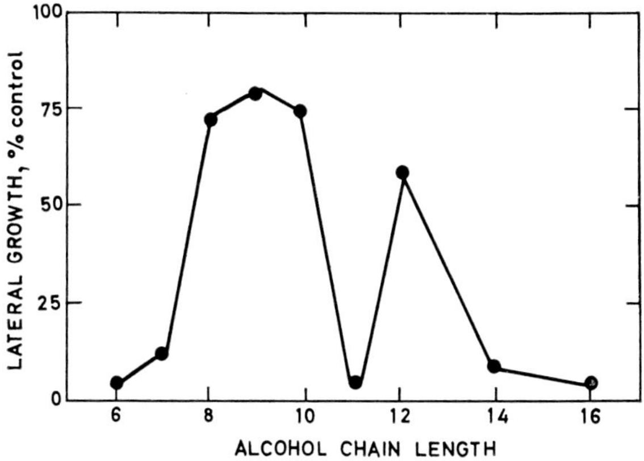

FIG. 6. *Axillary bud growth of 'Improved Indianapolis Yellow' chrysanthemum treated with fatty alcohols*·
21 days after treatment with 0·08M fatty alcohols and emulsified with 1·5% Tween-80
Control: manually pruned=100%

fatty alcohols killed only the terminal meristem. Axillary buds on treated plants grew at nearly the same rate as those on plants pruned manually. The C_{16} emulsion left a white residue on the foliage without apparent injury.

The propargyl compounds were compared at equimolar concentrations (Table I). The n-octyl compound was the only active structure at 2×10^{-2}M. The 2-octyl and n-octyl were active at 3×10^{-2}M. Regrowth of the axillary buds was similar on both groups of plants. The n-octyl emulsion caused much more visible damage to the mature leaves than did the 2-octyl emulsion. Compounds with longer or shorter alkyl chains than 8 carbons were inactive. At the concentration of 4×10^{-2}M, n-octyl, n-nonyl, and n-decyl killed the terminal meristem and partially damaged the axillary buds. Regrowth of the lateral branches of the treated plants was much less than the regrowth of the manually pruned plants. The 2-octyl emulsion at 4×10^{-2}M was almost as effective as the one at 3×10^{-2}M; the mature leaves were unaffected. The terminal meristem became flaccid 1 to 2 days after treatment with the propargyl alkyl sulphites and tan within 3 to 5 days.

Propargyl-n-decyl sulphite was compared with propargyl decanoate and propyl decanoates at equimolar concentrations of 3×10^{-2}, 4×10^{-2}, and 5×10^{-2}M. Propargyl-n-decyl sulphite was active only at a concentration of 5×10^{-2}M. It caused a great deal

Table I

Axillary bud growth (% control) of 'Improved Indianapolis Yellow' chrysanthemum* 21 days after treatment with various emulsions of alkyl propargyl sulphites further emulsified with 1% Atlox 3403

Compound of general formula R-OS(=O)-$OCH_2C{\equiv}CH$	Emulsion concentration			
	2×10^{-2}M	3×10^{-2}M	4×10^{-2}M	5×10^{-2}M
R=n-hexyl	0	0	0	—
n-octyl	96·8	97·2	28·4	—
n-nonyl	0	47·5	67·4	—
n-decyl	0	0	28·4	4·6
n-dodecyl	0	0	0	—
2-octyl	0	98·4	87·4	46·3
Propargyl decanoate	—	—	92·4	87·4
Propyl decanoate	—	—	0	0

*Weight of axillary branches expressed as % of weight of axillary branches of manually pruned plants

of leaf damage and inhibited axillary bud growth. Propargyl decanoate was active at a concentration of 4×10^{-2}M without any visible leaf damage. There was slightly less regrowth with a 5×10^{-2} than from a 4×10^{-2}M emulsion. Terminal meristems turned black within 15 minutes and flaccid in 30 to 60 minutes. Propyl decanoate was inactive at the dosages tested (4×10^{-2}, 5×10^{-2}M); it was active at 0·08 and 0·16M.

Similar results in response to fatty acid derivatives were observed with the other plants tested. The C_8 to C_{10} fatty acid esters, the C_8 to C_{11} fatty alcohols, the propargyl compounds (as propargyl-2-octyl sulphite and propargyl decanoate) were the most active compounds. Shorter and longer carbon chain compounds than C_8 to C_{10} were inactive.

Discussion

The most active fatty acid derivatives for the killing of the meristematic tissues were those with carbon chain lengths of 8, 9, 10, and 11 on all plant species tested.[4, 5, 8] On an equal molar basis, the ranking of activity of the three families of chemical pruning agents is: propargyl compounds>fatty alcohols>fatty acid esters. The response curves of all three families show that the test chemicals were either highly active or nearly inactive. The cut-off point between activity and no activity was usually ±2 carbons.

Emulsions of only the straight chain compounds were active. Branching of the ester moiety or the addition of a second methyl, ethyl, or propyl radical rendered the compounds inactive even when the basic alkyl group was 8 to 11 carbons in length. The total number of carbons in the compound did not determine activity; n-pentyl pentanoate (C_5–C_5) was inactive while C_8 either as n-octyl acetate or as ethyl octanoate was active. The fatty alcohols exhibited the same pattern. The most active compound was the C_{11} alcohol. The fatty alcohols also readily damaged the axillary buds; regrowth of the lateral branches was less than that observed on plants sprayed with fatty acid esters or pruned manually. The propargyl compounds were the most active of the fatty compounds tested and in this family, propargyl-n-octyl sulphite was the most active structure.

Structurally, all three families have a zigzag configuration with similar spatial dimensions: fatty acid ester—methyl decanoate (C_1+C_{10}); fatty alcohol—1-undecanol (C_{11}); propargyl compound—propargyl-2-octyl sulphite (C_3+C_8). These results suggest that there is a critical configuration for the fatty acid derivatives to exert their maximum effectiveness. Compounds containing a greater number of carbons than the optimal structures (C_{11}) caused a glazing and left a white deposit on the mature foliage. The leaf cuticle is composed of waxes which are short chain fatty esters and alcohols ranging in carbon chain length from C_{12} to C_{20}.[9] The most active structures, thus, have carbon chain lengths one to two carbons less than the compounds found occurring in the lipoidal film of the mature leaves. This lipoidal film is not yet present on immature leaves; the partially expanded epidermal cells in chrysanthemum leaves are protected only by the extensive occurrence of trichomes. The penetration of the fatty acid derivative–surfactant system thus occurs only with a liquid film of high wetting and penetrating power. Specific types and levels of surfactants produced the proper chemical and physical effects to enhance penetration and selectivity of various herbicides.[10] Surfactants in these present experiments were used not only to emulsify the fatty soluble compounds into water but also to limit the undesirable side effects of leaf burn and to reduce the amount of tissue killed to a minimum. The roles of surfactant type and amount will be published elsewhere.

It has been shown[8] that nuclear membranes of tobacco buds are ruptured by methyl decanoate emulsions. The permeability and/or structure of cytoplasmic membranes as well as membranes surrounding other cellular components could also be altered by the chemical. Substances such as phenols could then enter the cytoplasm in abnormally large quantities from the vacuole.[11] The responding areas turn black as a result of oxidised polyphenols which uncouple oxidative phosphorylation in the dying tissue. The changes in the properties of the membrane could result from a lack of sufficient energy for maintenance of their normal states. Highly vacuolate cells turned black initially, but often became green again several days after treatment. The cells without well-developed vacuoles, once they turned black, never recovered. Experimental evidence by the authors tends to support the proposition that the fatty acid derivatives act as uncouplers of oxidative phosphorylation as reported by Pedersen *et al.*[12] Changes in pH, for example, were detected but were measured only after visible damage had occurred. The authors have observed that the epidermal layer of chrysanthemum cells is damaged in the first minutes after treatment. Techniques sensitive enough to measure changes in acidity or release of polyphenols in the epidermal layer have not been developed.

Acknowledgments

The authors appreciate the co-operation of Atlas Chemical Industries, Inc., Wilmington, Del.; Emery Industries, Inc., Cincinnati, Ohio; the Procter & Gamble Co., Cincinnati, Ohio; and UniRoyal, Inc., Bethany, Conn., who supplied some of the chemicals. The chrysanthemum cuttings were supplied by Yoder Brothers, Inc., Barberton, Ohio. The top-loading balance used to weigh the lateral growth was provided by a grant from the Fred C. Gloeckner Foundation, New York.

References

[1] Le Poidevin, N., *Phytochem.*, 1956, **4,** 525
[2] Nieman, C., *Bact. Rev.*, 1954, **18,** 147
[3] Stowe, B. B., *Pl. Physiol., Lancaster*, 1960, **35,** 262
[4] Tso, T. C., *Nature, Lond.*, 1964, **202,** 511
[5] Tso, T. C., Steffens, G. L., & Englehaupt, M. E., *J. Agric. Fd. Chem.*, 1965, **13,** 78
[6] Tso, T. C., Burk, L. G., & Steffens, G. L., *Tobacco, N.Y.*, 1966, **163,** 26
[7] Cathey, H. M., Steffens, G. L., Stuart, N. W., & Zimmerman, R. H., *Science, N.Y.*, 1966, **153,** 1382
[8] Steffens, G. L., Tso, T. C., & Spaulding, D. W., *J. Agric. Fd. Chem.*, 1967, **15,** 972
[9] Stumpf, P. K., 'Plant Biochemistry', 1965, p. 324 (New York and London: Academic Press)
[10] Jansen, L. L., Gentner, W. A., & Shaw, W. C., *Weeds*, 1961, **9,** 381
[11] Spencer, M., 'Plant Biochemistry', 1965, p. 802 (New York and London: Academic Press)
[12] Pedersen, T. A., Kirk, M., & Bassham, J. A., *Biochim. biophys. Acta*, 1966, **112,** 189

DISCUSSION

Dr. C. R. Worthing.—It is well known that the symptoms and incidence of phytotoxicity vary profoundly under different environmental conditions, e.g. stage of growth of plant, temperature, sunlight, humidity, etc. Do you find the same variation with your 'growth assassins', as this might complicate their use commercially?

Dr. H. M. Cathey.—The use of chemical 'growth assassins' is complicated by the environment. The chemicals are emulsified with surfactants which regulate their action and safen the response. We are finding that they can be used to regulate growth successfully after one learns to handle them.

Dr. C. R. Worthing.—Are any of the surfactants active as assassins *per se*?

Dr. H. M. Cathey.—The surfactants we are using are inert in so far as we can observe. Many types have been tried, only a few set up the proper functioning of the chemicals.

Prof. B. B. Stowe.—It is curious that the chain lengths effective in your system reach an upper limit at C_{12}. This is exactly where growth promotion in our excised pea stem sections by fatty acid esters begins. Have you ever noted growth stimulations, however small, by your compounds?

Dr. H. M. Cathey.—We have not observed any growth stimulation of plants treated with carbon chain lengths of 8 to 12. They may be very toxic unless emulsified with the proper kind and amount of surfactant. We then observe very selective kill of the vegetative tip without damage to the remaining parts of the plant.

Dr. B. J. Heywood.—Do molecules of similar size, shape and polarity to the C_{10} methyl esters, e.g. acetamidodecane, have this property?

Dr. H. M. Cathey.—We have tested many compounds as chemical pruning agents. The active compounds were fatty acid esters or alcohols. When other radicals were combined with the aliphatic chain the compounds were usually inactive. We have not tested acetamidodecane.

Dr. J. R. Corbett.—If selective killing of particular buds is due solely to the surfactant, could you use dinitro-*o*-cresol or other standard herbicides, instead of petroleum products?

Dr. H. M. Cathey.—We have a number of different standard herbicides in our chemical disbudding test on chrysanthemums. They inhibit or kill all plant parts; we could not identify any selectivity of action between the terminal and lateral flower buds. Most simple aromatic compounds were effective disbudding agents when emulsified with the proper surfactant.

ETHYLENE AS A PLANT HORMONE

By DAPHNE J. OSBORNE

(A.R.C. Unit of Experimental Agronomy, Department of Agriculture, University of Oxford)

The effect of ethylene on plant growth is reviewed. Not only are the growth responses of certain tissues closely linked to their ethylene production and can be evoked by suitable concentrations of externally applied ethylene, but evidence is presented from experiments on auxin-induced defoliation that ethylene produced in one tissue may regulate growth and senescence in an adjacent tissue. In etiolated pea stems, ethylene changes not only the growth, but also the enzyme content of treated tissues, and experiments are described that suggest that the synthesis of peroxidase may be regulated by ethylene. Ethylene appears, therefore, to possess the general characteristics of a plant hormone.

Introduction

THE interest in ethylene as a plant growth regulating chemical has extended over many years and although it was clear even in the middle 1800's that the presence of gaseous materials in the air could modify the growth of plants,[1] the identity of ethylene as an active agent was not established until the turn of the century.

Flower growers, for example, knew that traces of illuminating gas escaping in their glass houses would result in premature closure of the blossoms,[2] while Philippine mango growers and Puerto Rican pineapple farmers used to light bonfires near their fields believing that the smoke helped to initiate and synchronise the flowering of their crops.[3] Both of these effects we now attribute to the action of ethylene.

It was Neljubow[4] who in 1901 reported that the concentrations of ethylene that occur in illuminating gas caused curvature responses, stem thickening and loss of response to gravity in pea stems. Low concentrations produced only growth inhibition, and when the ethylene was removed, growth continued normally again. Since these effects were essentially similar to those observed in pea plants exposed to illuminating gas, he concluded that ethylene could be the growth regulating constituent.

Later Molisch,[5] in 1916, showed that smoke produced by burning paper would cause similar kinds of growth responses in plants, and again, ethylene was indicated as the causative agent in the vapour.

From 1901 to the 1920s many people exposed plants to the vapours of unsaturated hydrocarbons, and the experimental records of this work make fascinating reading today (e.g. *Contr. Boyce Thompson Inst. Pl. Res.*, 1908–1935). Two major points emerge from these studies. Firstly, it is clear that ethylene itself is by far the most effective hydrocarbon, being 100–1000 times more active than related chemicals: inhibitions of stem growth that occur at very low concentration (0·02–1 ppm) are, in fact, specific to ethylene. Secondly, plants vary considerably in their sensitivity to ethylene in the air. African marigolds show curvature responses to as little as 0·001 ppm,[5] while the minimum concentration reported for responses of rose plants is 0·3 ppm.[6] Sensitivity also varies with the age of the tissue.[5]

The major growth responses to ethylene are: (1) inhibition of growth by cell extension, (2) acceleration of abscission of leaves, fruits and flowers, (3) ripening of fruits, (4) epinasty of leaves, (5) loss of response to gravity and light, (6) breakage of bud dormancy, (7) stimulation of cell division in meristematic cells of cambium and lenticels and (8) enhancement of flowering. The first three only are considered here.

Production of ethylene by plants

The first indication that ethylene was a natural plant product came in 1932, when Elmer[7] found that apples gave off a volatile substance that inhibited the growth of potato sprouts enclosed with them.

A year later, Botjes[8] showed that a volatile product from apples causing epinasty in tomato seedlings was removed from the air by ethylene absorbents. From this and later work of Gane[9] and others, ethylene was established as a natural plant product with growth regulating properties.

Not all parts of a plant were found to produce the same amount of ethylene. Young leaves of vegetative plants, for example, produce more than old. Rates of 1–3 mμl/g/h have been recorded for young cotton leaves compared with a detectable trace only by senescent leaves.[10,11] Experiments in these laboratories using *Xanthium pennsylvanicum*, in which ethylene collections were made over 24 h from detached leaves from nodes 2–10 below the apex, gave decreasing amounts of ethylene from leaves of increasing age.

In fruits, however, the situation is reversed. Young fruits produce less ethylene than ripe fruits. Analysis of gas samples withdrawn from the central cavity of melons[12] has shown an internal ethylene concentration of 0·04 ppm at ten days after pollination, rising to 10–70 ppm just prior to the climacteric—an increase of 1000 fold. Flowers also produce measurable quantities of ethylene.[13,14] For Vanda orchids Burg & Dijkman[15] recorded rates of 0·35 mμl/g/h following pollination, but no ethylene production was observed in unpollinated controls.

There appears to be difference in behaviour between leaves, and fruits or flowers in their response to applied ethylene.

Exposure for a few hours to ethylene causes the onset of ripening in mature fruits and triggers the natural rise in ethylene production that precedes the normal process of fruit ripening. This has been clearly demonstrated for bananas,[16] where the applied ethylene apparently functions as an autocatalyst for ethylene biosynthesis. For many years it has been a common commercial practice to use ethylene to ripen fruit such as bananas, pineapples and citrus that are harvested while still green. Because of this regulatory role in the ripening process, ethylene has often been referred to as the 'ripening hormone'.[17]

Autocatalytic ethylene production is implied in Vanda orchid flowers[15] where exposure of blossoms to 10 ppm ethylene in air stimulated ethylene production particularly by the column and led to the premature fading of the flower. But a different mechanism appears to operate in the ethylene induced senescence and abscission of leaves. Ethylene production appears to be minimal when the leaf is senescent,[11] and although ethylene hastens yellowing and subsequent leaf fall, there is no evidence to suggest that ethylene has an autocatalytic effect on ethylene synthesis by leaf tissue.

Control of ethylene production and its relation to growth

Zimmerman & Wilcoxon[18] and later Hansen[19] in 1946 provided the earliest reports that the production of ethylene might be controlled by other plant growth regulators. Hansen's experiments, for example, indicated that the ethylene production of pear fruits was enhanced by treatment with the synthetic auxin 2,4-dichlorophenoxyacetic acid (2,4-D). But Morgan & Hall[20,21] were the first to demonstrate unequivocally that the ethylene production of leaf and stem tissue was increased many fold after spraying plants with 2,4-D or with the naturally occurring auxin indole-3-acetic acid.

Abeles & Rubinstein[22] subsequently showed that the lateral redistribution of endogenous auxin in hypocotyls of *Phaseolus* seedlings following a phototropic or geotropic stimulus also results in a greater evolution of ethylene by the half of the hypocotyl that is auxin-enriched e.g. the darkened half in unilaterally illuminated seedlings and the lower half in seedlings placed in a horizontal position. In addition, they showed that applied gibberellin as well as applied auxin will stimulate ethylene formation in stem tissue of *Phaseolus* while treatment with a number of growth retardants will reduce the normal level of production.

Although ethylene is a natural plant product, and externally applied ethylene has profound effects on plant growth, its place as a natural plant hormone is still debated.[23] Much of the evidence available indicates that ethylene could be a product of fruit ripening and flower fading, and that in vegetative parts the production of ethylene can be causally linked with the auxin content of the tissue.

The results of a number of studies with etiolated seedlings of *Pisum* and *Phaseolus* suggest that ethylene exerts considerable control of the growth of cells in which it is produced. For example, Burg & Burg[24] showed that, in segments of pea stems, auxin concentrations which stimulated ethylene production were those at which auxin was supraoptimal for maximal cell extension, and the reduction in growth at high auxin concentrations could partly be accounted for by the growth inhibitory effects of the ethylene produced.

However, the clearest example of the correlation between ethylene production and cell extension has been provided by studies of the growth and uncurling of the plumular hook in etiolated seedlings of *Pisum* and *Phaseolus*. When seedlings are grown in darkness the plumule is curled over at an angle of 180°, but on receiving white or red light, growth of the hook and plumule is enhanced, the hook uncurls and the shoots attain a vertical position. Kang, Yocum, Burg & Ray[25] showed that the light-induced opening in *Phaseolus* could be prevented by 0·1–0·5 ppm ethylene in the air. Precise measurements of growth of hooks and natural ethylene production in *Pisum* seedlings have been made by Goeschl, Pratt & Bonner.[26] They found that ethylene production of the shoot was confined to the plumule, the plumular hook and the adjacent 5 mm of epicotyl. After exposing the seedlings to 60 seconds of red light and returning them to darkness the ethylene production decreased to a minimum after 9 h and then rose again, while growth of plumule rose to a maximum rate and then declined over the same period. Growth could be closely correlated with the natural rate of ethylene production.

The growth and uncurling of the plumular hook is known to be controlled by phytochrome; the uncurling effect of red light being reversed when followed by far red light. Measurements of ethylene production after this light sequence in *Pisum*[26] indicated that more ethylene was formed than in seedlings exposed only to red light but the level did not rise to that of the dark controls. Unfortunately the far red source used by Goeschl, Pratt & Bonner contained some red wavelengths so that complete reversal of the red effect by far red was not shown. However, despite this, the results indicate that ethylene production might be geared to the phytochrome regulatory system. These results, and those showing that the application of ethylene on concentrations of auxin that result in enhanced ethylene production,[25] will prevent light-induced hook opening, suggest that ethylene does exert a regulatory role in the growth of the hook cells in which it is produced, and that it is not merely a byproduct of the growth response.

Evidence for hormonal activity

Having considered the regulatory role of ethylene in tissues in which it is produced, it is pertinent to consider the requirements for its being classified as a hormone. By classical definition a hormone is a substance that is produced at one site and is transported to another site where, in very small concentrations, it exerts a regulatory effect upon the growth of the target tissue.

Evidence is scant that the ethylene produced by one part of a plant is transported to other tissues and there controls growth processes. Hall, Miller & Herrero[27] found that radioactivity from ^{14}C-ethylene supplied to a single attached leaf of cotton plants was found in at least 18 different metabolites, in other parts of the plant, 15 h later. Some of the radioactivity was possibly transported as unchanged ethylene. The studies described below on the regulation of leaf senescence and abscission have added further information on the role of ethylene and indicate that the local movement of ethylene in plants may be of considerable importance in the control of growth processes.

Experimental

These investigations have centred on the defoliant action of aerial sprays of synthetic auxins in temperate and topical evergreens. The n-butyl ester of 2,4-D is remarkably effective in this respect and formulations containing these auxins are usual defoliants in East Africa, Malaya and Vietnam. In England, abscission of second year leaves of *Euonymus japonica* can be affected in 2 weeks at 24° by applying drops of the ester in methanol to the surface of the blade. The treated area remains green, but the parts that surround the treated spot show visible signs of yellowing after 4 days. Yellowing proceeds centrifugally from the spots and, at abscission, the whole of the untreated part of the blade is either yellow or partly senescent[28] (Fig. 1). Naturally senescent leaves[29] and leaves in which senescence is induced by auxins[30] contain abscission accelerating substances which will trigger the biochemical processes that lead to cell separation at the abscission zone. But the mechanism by which the auxin treatment causes premature senescence in the untreated part of the leaf blade has been less clear. Evidence suggested that in competition for nutrients, the enhanced metabolism of the auxin treated spots resulted in a depletion of substrates from the surrounding untreated parts of the blade[31] and a speeding of their senescence. Further studies indicate an additional explanation.

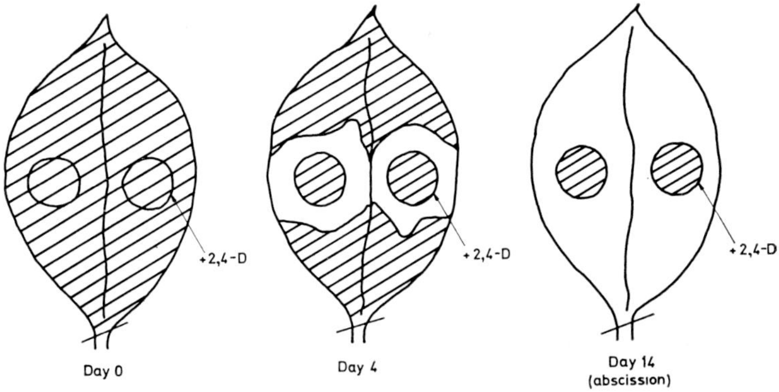

FIG. 1. *Development of senescence in leaf blade of* Euonymus japonica *following treatment with* n-*butyl ester of* 2,4-*dichlorophenoxyacetic acid*
Circled region treated with ester, 6·66 mg/ml in methanol
▨ green ☐ yellow and senescent

Experiments in which the ethylene was separately collected from control halves and 2,4-D treated halves of *Euonymus* leaves over periods of 3 days show that the auxin greatly increases the production of the gas over a prolonged period. Methods of ethylene collection were similar to those described by Young, Pratt & Biale[32] and ethylene was subsequently determined in an F & M Gas Chromatograph against suitable standards. Typical results are shown in Table I. If branches of *Euonymus* are exposed to ethylene at 5–300 ppm, loss of chlorophyll and leaf fall occur

Table I

Production of ethylene by leaves of Euonymus japonica *treated with 2,4-D*

Collection period, h	C_2H_4 production, mμl/g/h	
	Control*	With 2,4-D†
0–24	1·2	15·1
24–47	3·4	14·8
47–71	4·4	22·8

*Halved leaves of *Euonymus japonica* treated on the adaxial surface with methanol

†Halved leaves of *Euonymus japonica* treated on the adaxial surface with 2,4-D butyl ester 6·66 mg/ml methanol

within 2 weeks at 24°. Detached leaves show visible signs of senescence after 4 days if they are enclosed with 2,4-D treated leaves or in air containing 5–300 ppm of ethylene. No senescence occurs when they are enclosed for this time with untreated leaves or in air.

These results suggest that the ethylene produced by the auxin-treated part of the leaf is instrumental in causing premature senescence of the surrounding cells and that senescence follows the centrifugal movement of ethylene from the treated areas. The absence of senescence in the 2,4-D treated parts would appear to result from the protection that the high concentration of auxin affords against the effects of this ethylene for using the n-butyl ester of ^{14}C-2,4-D it was shown that the ^{14}C was retained within the treated area.[31] Also, branches whose leaves are treated over the whole of the adaxial surface with 2,4-D ester show a delay of two weeks or more compared with controls in the appearance of yellowing and eventual leaf abscission when both kinds of branches are exposed to 500 ppm ethylene. Auxin-ethylene interactions of this general type have been reported before in a number of plant tissues.[33,34]

Although auxin-induced ethylene production appears to be the causative agent responsible for leaf senescence and abscission in *Euonymus,* it is not clear whether the ethylene passes from the auxin-treated to the untreated tissues in the gas phase, or whether it moves in solution from cell to cell. If it moves in the gas phase, both inside and outside the leaf, ethylene should perhaps be considered as the equivalent of an insect pheromone. But since the definition of a hormone does not define the path of movement it is proposed that the auxin-induced defoliation described here is evidence for the hormonal regulation of leaf senescence in one part of a leaf blade by ethylene produced in, and moving from, an adjacent region.[28]

In a modern interpretation of hormonal action the hormone is expected to regulate not only the visible growth and development of the plant but also the pattern of enzyme differentiation. Such effects of hormones are, in most instances, attributable to a regulation of the synthesis of new types of protein. Examples are the production of α-amylase in the barley aleurone[35] and invertase in sugar beet storage tissue[36] as a response to gibberellic acid; the enhancement of cellulase in pea epicotyls by auxin would seem to be another example.[37]

Several reports have appeared in the past to indicate that ethylene exerts a regulatory role in enzyme activity. Ivanov,[38] Hansen[39] and later Herrero & Hall[40] found decreases in catalase and increases in hydrolytic enzymes and peroxidase as a result of ethylene treatments at 1000 ppm.

More recent studies in these laboratories on peroxidase in pea stems indicate that ethylene, like auxin and gibberellin, will modify enzyme activity and regulate nucleic acid and protein synthesis. The material used in these studies has been dark grown seedlings of *Pisum sativum* c.v. 'Alaska'. They were used seven days after planting, when the third internode had commenced extension. Peroxidase was extracted by homogenising the tissue in 0·005M potassium maleate buffer at pH 5·8, and after centrifuging at 2500 *g*, the enzyme activity was determined by a modification of the method described by Ponting & Josslyn,[41] and is expressed in arbitrary units based on the rate of formation of the coloured oxidation products of guaiacol in the presence of hydrogen peroxide.

If etiolated pea plants are exposed to ethylene for 24 h at concentrations from 1–500 ppm, the extension growth of the apical 5 mm of the epicotyl is inhibited. At 500 ppm (Fig. 2), extension growth in this region of the intact shoot is reduced to 3% of the air control, but the cells in the treated plants increase in size laterally; the increase in fresh weight of this region is 19% of the control. This change in growth pattern in response to ethylene also occurs in excised segments. When they are floated on solutions of maleate buffer at pH 5·8 containing 2% sucrose and exposed to air or ethylene for 24 h, treated segments extend less in length than controls in air, but their increase in fresh weight may actually be greater.

The reduction of cell elongation in ethylene, and the enhance-

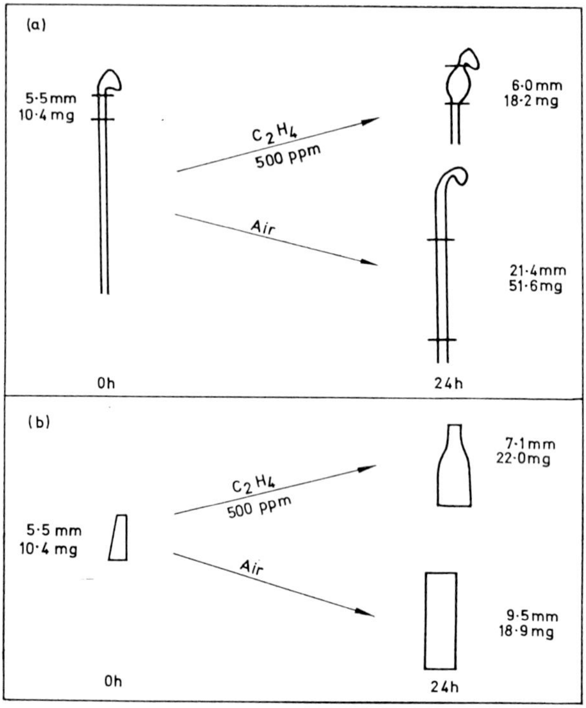

FIG. 2. *Mean values for extension growth and fresh weight of* Pisum sativum *after* 24 *h in air or* 500 *ppm ethylene*
(a) marked 5 mm segments from the apex of intact etiolated *Pisum* seedlings
(b) 5 mm segments excised from this region and floated on potassium maleate buffer pH 5·8 containing 2% sucrose

ment of lateral expansion, is confined to the extending regions of the pea plant; no effects on growth or uptake of water are discernible in the fully extended parts of mature internodes. Although measurements of peroxidase activity in extending and mature internodes show that ethylene markedly stimulates the activity of this enzyme in all parts of the epicotyl, only those results pertaining to the apical 5 mm of extending third internodes will be discussed here.

A direct comparison of the peroxidase activity in the apical region of intact seedlings and in excised segments following 24 h in ethylene 500 ppm is shown in Fig. 3, and it is seen that significant increases occur in the tissues under both conditions. Time course studies show that there is a lag in the peroxidase rise in response to ethylene lasting at least 2 h, but this is followed by a

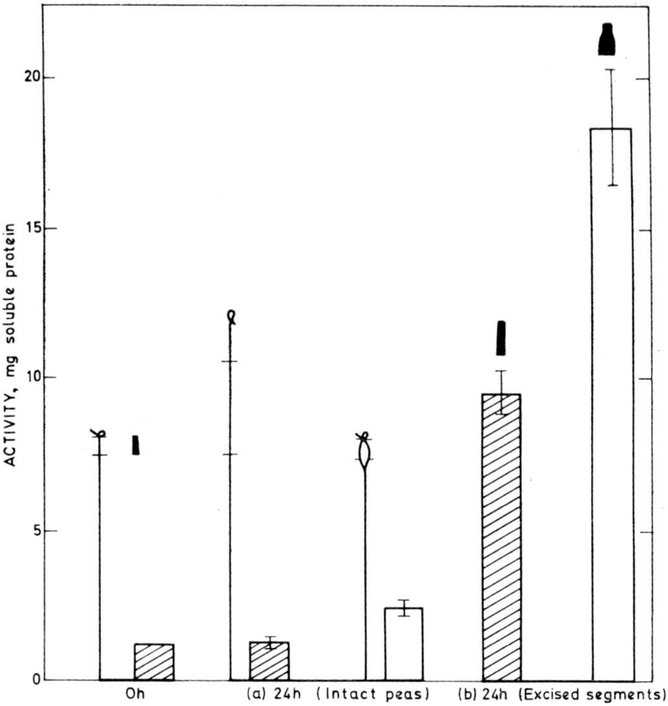

FIG. 3. *Relative peroxidase activities of* Pisum sativum *apical tissue after* 24 *h in air or* 500 *ppm ethylene*

(a) marked 5 mm segments from the apex of etiolated *Pisum* seedlings

(b) 5 mm segments excised from this region

▨ air □ C_2H_4 500 ppm

stimulation that continues for a further 24 h. The rise in peroxidase activity in segments of pea stem following excision is already well known, and has been shown to include the appearance of a new iso-enzyme.[42] The present experiments carried out in sterile conditions show that the ability of apical segments to incorporate radioactively labelled amino acids into protein is significantly enhanced over control levels when segments are pretreated with ethylene 500 ppm for 24 h (Table II). This indicates that the ethylene enhanced peroxidase activity in the segments is associated with an ethylene enhanced synthesis of protein.

If inhibitors of ribonucleic acid and protein synthesis are included in the medium when segments are exposed to air or ethylene for 24 h (Fig. 4), it is found that both the natural and the

Table II

Uptake and incorporation of ^{14}C-leucine into protein in apical segments of Pisum sativum

Apical segments	cpm	
	Relative uptake	mg protein
0 h	261 ± 22	873 ± 8
24 h air	1465 ± 45	4885 ± 151
24 h C_2H_4	1544 ± 47	5570 ± 82

At 0 h and after 24 h in air or C_2H_4 500 ppm. Incorporation 2 h. For methods see.[28]

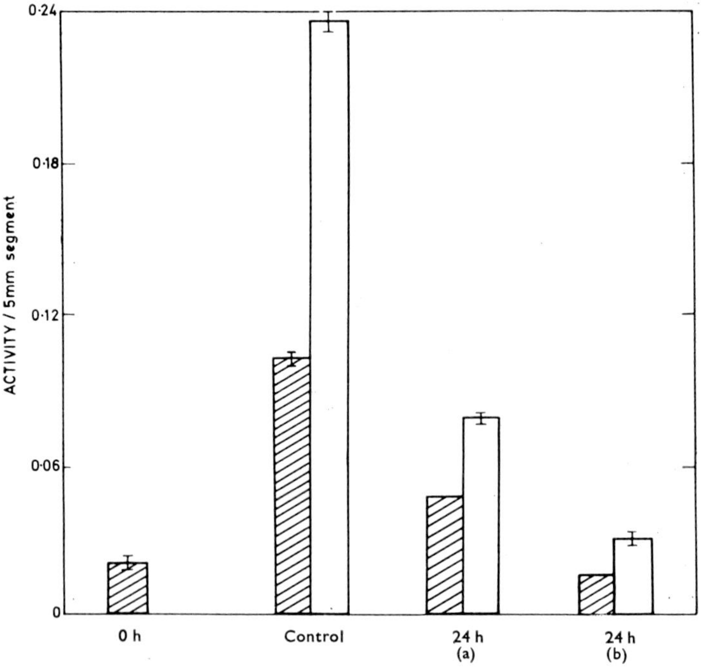

FIG. 4. *Relative peroxidase activities in 5 mm segments excised from the apex of etiolated* Pisum sativum *seedlings*
(a) cycloheximide 3 mg/l, or
(b) actinomycin D 30 mg/l included in the basal medium, during exposure to air or ethylene 24 h

/// air ☐ C_2H_4 500 ppm

ethylene-stimulated rise in peroxidase activity is inhibited. Actinomycin D at 30 mg/l gives almost complete inhibition, and cycloheximide 3 mg/l permits less than 40% of the control rise. Higher concentrations of cycloheximide reduce this level further.

However, if the segments are transferred to solutions of the inhibitors 11 h after exposure to air or ethylene (Fig. 5), cycloheximide reduces the further rise in enzyme activity whereas

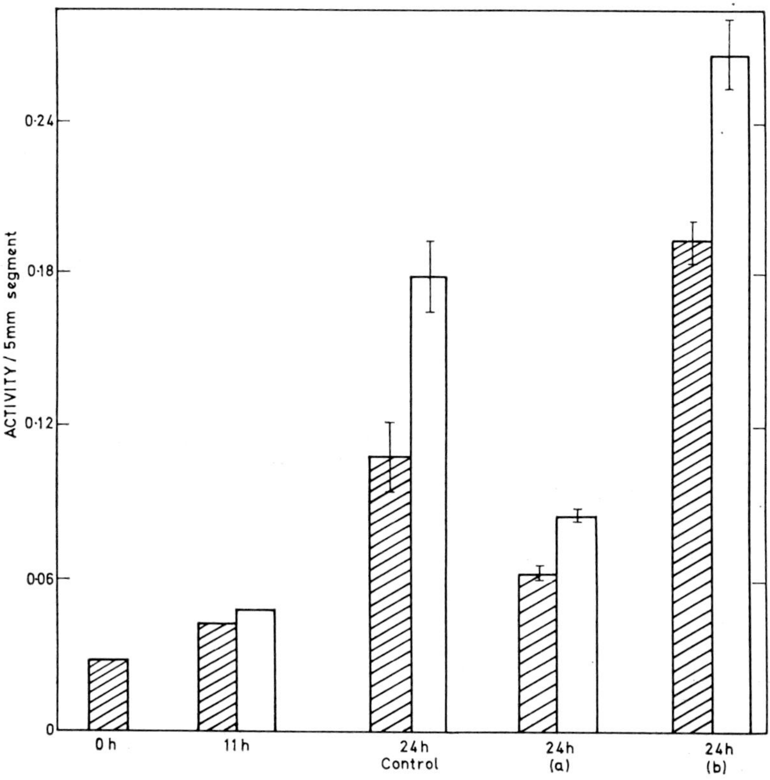

FIG. 5. *Relative peroxidase activities in 5 mm segments excised from the apex of etiolated* Pisum sativum *seedlings after treatment air or ethylene for 11 h*

(a) subsequently transferred to cycloheximide 6 mg/l for a further 24 h in air

(b) subsequently transferred to actinomycin D 10 mg/l for a further 24 h in air.

▨ air □ C_2H_4 500 ppm

actinomycin treatment has no inhibitory effect. This suggests that both RNA and protein synthesis are necessary to permit the initiation of the rise in peroxidase in both air and ethylene, but that once the rise has been initiated, inhibition of protein synthesis, but not RNA synthesis, will prevent the further increase that occurs with time. This would accord with the concept that a stable mRNA for peroxidase had become available during the 11 h pretreatment period.

It may be noted that the addition of actinomycin after 11 h results in a higher activity than the control value after a further 24 h. This could perhaps result from the actinomycin inhibition of the synthesis of an inhibitor of the enzyme. Inhibitors of peroxidases are present in pea stem tissue and the possibility that part of the increase in peroxidase activity in ethylene could be due to the enhanced leakage or destruction of such an inhibitor, cannot be ignored. There is the further possibility that the increase in peroxidase activity in the presence of ethylene could be attributed to a reduction in the leakage of the enzyme into the surrounding medium. Although ethylene does slightly reduce the peroxidase activity of the medium compared with that of controls in air, the effect can account for less than 1% of the increase in the tissue. Although such calculations are necessarily complicated by possible differential effects of the treatments on leakage and stability of inhibitors, it seems unlikely that the whole of ethylene enhanced peroxidase activity in excised segments or intact plants can be accounted for in this way. Clearly, ethylene is a factor in the regulation of peroxidase activity in pea stem tissue, and the enhanced activity due to ethylene is subject to inhibition by the usual inhibitors of RNA and protein synthesis. In this respect, ethylene would appear to act in a way similar to that described for other naturally occurring plant hormones in the control of enzyme activities.

Conclusion

The kinds of biological effects of ethylene listed in this survey are evidence that ethylene should be included among the natural plant hormones.

Acknowledgments

The author wishes to acknowledge the co-operation of Dr. Mary Hallaway and Mrs. Irene Ridge in experimental work included in this paper.

References

[1] Girardin, J. P., *Jber. AgrkChem.*, 1864, **7,** 199
[2] Knight, L. T., & Crocker, W., *Bot. Gaz.*, 1913, **55,** 337
[3] Rodrigues, A. R., *J. Puerto Rico Dep. Agric.*, 1932, **16,** 5
[4] Neljubow, D., *Bot. Centralbl.*, 1901, **10,** 128
[5] Crocker, W., 'Growth of plants', 1948 (New York: Reinhold)
[6] Zimmerman, P. W., Hitchcock, A. E., & Crocker, W., *Contr. Boyce Thompson Inst. Pl. Res.*, 1931, **3,** 459
[7] Elmer, O. H., *Science, N.Y.*, 1932, **75,** 193
[8] Botjes, J. O., *Tijdschr. Plantenziekten*, 1933, **39,** 207
[9] Gane, R., *Nature, Lond.*, 1934, **134,** 1008
[10] Hall, W. C., *Bot. Gaz.*, 1951, **113,** 55
[11] Hall, W. C., Truchelut, G. B., Leinweber, L. L., & Herrero, F. A., *Physiologia Pl.*, 1957, **10,** 306
[12] Lyons, J. M., McGlasson, W. B., & Pratt, H. K., *Pl. Physiol., Lancaster*, 1962, **37,** 31
[13] Fischer, C. W., *St. Flower Growers Bull.*, 1950, **61,** 1
[14] Ton, Phan-Chon, *C.R. hebd. Seanc. Acad. Agric. Fr.*, 1963, **9**
[15] Burg, S., & Dijkman, M. J., *Pl. Physiol., Lancaster*, 1967, **42,** 1648
[16] Mapson, L. W., & Robinson, J. E., *J. Fd Technol.*, 1966, **1,** 215
[17] Burg, S. P., & Burg, E. A., *Science, N.Y.*, 1965, **148,** 1190
[18] Zimmerman, P. W., & Wilcoxon, F., *Contr. Boyce Thompson Inst. Pl. Res.*, 1935, **7,** 209
[19] Hansen, E., *Pl. Physiol., Lancaster*, 1946, **21,** 588
[20] Morgan, P. W., & Hall, W. C., *Physiologia Pl.*, 1962, **15,** 420
[21] Morgan, P. W., & Hall, W. C., *Nature, Lond.*, 1964, **201,** 99
[22] Abeles, F. B., & Rubinstein, B., *Pl. Physiol., Lancaster*, 1964, **39,** 963
[23] Osborne, D. J., 'Biochemistry and physiology of plant growth substances', (Ottawa: Range Press) in press
[24] Burg, S. P., & Burg, E. A., *Proc. nat. Acad. Sci., U.S.A.*, 1966, **55,** 262
[25] Kang, B. G., Yocum, L. S., Burg, S. P., & Ray, P. M., *Science, N.Y.*, 1967, **156,** 958
[26] Goeschl, J. D., Pratt, H. K., & Bonner, B. A., *Pl. Physiol., Lancaster*, 1967, **42,** 1077
[27] Hall, W. C., Miller, C. S., & Herrero, F. A., 'Plant growth regulation', 1961, p. 751 (Iowa: State Press) (Ed. Klein, R. M.)
[28] Osborne, D. J., *Symp. Soc. Exp. Biol.*, 1967, **21,** 305
[29] Osborne, D. J., *Nature, Lond.*, 1955, **176,** 1161
[30] Osborne, D. J., *Trop. Agric. Trin.*, 1958, **35,** 145
[31] Osborne, D. J., & Hallaway, H. M., 'Plant growth regulation', 1961, p. 329 (Iowa: State Press) (Ed. Klein, R. M.)
[32] Young, R. E., Pratt, H. K., & Biale, J. B., *Analyt. Chem.*, 1952, **24,** 551
[33] Gawadi, A. G., & Avery, G. S., *Am. J. Bot.*, 1950, **37,** 172
[34] Hall, W. C., *Bot. Gaz.*, 1952, **113,** 310
[35] Varner, J. E., & Chandra, G. R., *Proc. nat. Acad. Sci., U.S.A.*, 1964, **52,** 100
[36] Cherry, J. H., 'Biochemistry and physiology of plant growth substances', (Ottawa: Range Press) in press
[37] Fan, Der-Fong, & Maclachlan, G. A., *Can. J. Bot.*, 1966, **44,** 1025
[38] Ivanov, N. N., *Biochem. Z.*, 1932, **254,** 71
[39] Hansen, E., *Pl. Physiol., Lancaster*, 1939, **14,** 145
[40] Herrero, F. A., & Hall, W. C., *Physiologia Pl.*, 1960, **13,** 736
[41] Ponting, J. D., & Josslyn, M. A., *Arch. Biochem.*, 1948, **19,** 47
[42] Ockerse, R., Siegel, B. Z., & Galston, A. W., *Science, N.Y.*, 1966, **151,** 452

DISCUSSION

Mr. A. R. Sheldrake.—Does auxin have a senescence retarding effect on the evergreen leaves investigated? If not, why does the increased production of ethylene not cause the parts of the leaf which are producing it to turn yellow themselves?

Dr. Osborne.—Applied auxin does have a senescence retarding effect in *Euonymus* leaves exposed to ethylene. The evidence suggests that there is an auxin (applied+endogenous)/ethylene balance in favour of auxin in the treated part of the blade, but that the auxin (endogenous only)/ethylene balance in the untreated part of the blade lies in favour of the senescence accelerating effects of ethylene.

Dr. J. W. Dicks.—Is there any special significance attached to the choice of peroxidase in your investigations?

Dr. Osborne.—Yes. As you know peroxidase enzymes are implicated in the endogenous regulation of auxin levels and in the ageing and differentiation of plant cells. High activity is associated with lignification of vascular tissue, with senescent changes in leaves and with leaf abscission. Peroxidase has a special interest to us in relation to senescence and its possible role in cell extension.

Prof. B. B. Stowe.—The levels of 500 ppm ethylene used in the pea peroxidase experiments are high. Do the increases observed occur at lower levels?

Dr. Osborne.—These levels are high compared with the minimum concentration of ethylene required to produce growth inhibition. They are, however, quite non-toxic and plants recover and grow normally following exposure. Increases in peroxidase activity are detectable after 24 h in 1 ppm ethylene but are maximal at 100–500 ppm ethylene.

RESPONSES OF CROP PLANTS TO GROWTH REGULATORS

By E. C. HUMPHRIES

(Rothamsted Experimental Station, Harpenden, Herts.)

The general effects of the growth regulators, gibberellic acid, CCC [(2-chloroethyl)-trimethylammonium chloride] and B9 (dimethylaminosuccinamic acid) on crop plants are described. Gibberellic acid has not fulfilled its promise as a useful growth regulator but its full potential has probably yet to be realised.

Besides shortening stems CCC and B9 help to maintain the protein content of leaves, and CCC increases the rate at which leaves are produced in sugar beet. However, one of the important practical effects of CCC and B9 is to redistribute dry matter in the plant; in this way B9 increases the growth of potato tubers.

CCC has been widely used on cereals, particularly wheat, to prevent lodging but, in several experiments at Rothamsted and Woburn, CCC increased yield of crops that did not lodge. This seems to be because CCC enlarges the size of the root system and enables more tillers to survive at critical times. When CCC was applied at 2½ lb/acre small residues of an inhibitor (probably unchanged CCC) could be detected by the straight-growth test. This assay also showed that grain from CCC-treated crop contained less inhibitor when sprayed early than when sprayed late.

CCC has less effect on barley, oats and rye. At first the application of CCC shortens barley, but later the treated plants grow faster than untreated plants because the upper internodes elongate more.

Introduction

THE productivity of field crops so far has been increased by using varieties that give greater yields, by improved nutrition and by better control of weeds, pests and diseases but, in future, chemicals that control plant growth may also contribute to better yields. This could happen in two ways: (1) by adapting existing varieties to grow better in their environment and (2) by altering the relative proportion of different plant parts and increasing the yield of the economically important parts, e.g. grains of cereals and tubers of potatoes. The chemical control of growth in agricultural plants is only beginning but several chemicals fulfil the requirements mentioned. With more knowledge about the growth processes that contribute to yield, chemicals may be devised to control plant growth as desired.

This paper deals with some general responses of crop plants to growth regulators and gives examples of how they have been used in field crops.

A growth regulator is a substance that, applied in small amounts, alters the form of a plant by altering the relative proportions of its component parts. This is not a rigid definition because it includes substances that are not true growth regulators, but will do for the purposes of this paper in which discussion is confined to substances like gibberellic acid, CCC [(2-chloroethyl)-trimethylammonium chloride)] and B9 (dimethylaminosuccinamic acid), which either accelerate plant growth or interfere with the synthesis and metabolism of endogenous hormones and so alter plant habit or form.

Gibberellic acid

When the properties of gibberellic acid were first described they seemed to offer possibilities of increasing crop yield, but so far these have not been fulfilled. However, the use of gibberellic acid has yet to be fully explored, partly because its cost has prohibited extensive field trials. There is now renewed interest because unrefined gibberellin preparations may be used[1] to increase sugar yield from sugar cane.

Gibberellic acid is most likely to assist crop production by its property of stimulating leaf growth, which in turn increases leaf area and dry-matter production. It has been found[2] that spraying a 50 mg/l solution on to potato foliage increased area of leaves already actively growing and also of leaves not yet developed. These leaves, although paler than untreated leaves, were equally as efficient in photosynthesis and produced more dry matter. However, when the plants were mature they usually had produced only the same amount of dry matter as untreated plants, probably because soil volume and perhaps mineral nutrients were limiting. When the amount of soil per plant was increased, yields of tubers were increased by gibberellic acid.[3]

General effects of growth regulators

Shortening stems is the most characteristic effect of CCC. In small amounts, CCC does not affect leaf size but sometimes increases total leaf area. For example, in mustard (*Sinapis alba*) shortening the stem causes lateral stems to grow more, and total leaf area is increased.[4] The leaves are usually darker green and may persist longer. In dwarf French bean (*Phaseolus vulgaris*) CCC causes the primary leaves to remain green longer than do the leaves of untreated plants. The total protein of leaves of untreated plants declines sooner than in leaves of plants treated with CCC or B9, probably because the slower growing shoot of treated plants has a smaller demand for nitrogen.[5]

CCC also causes leaves of sugar beet to emerge faster,[6] but this has no practical value because the leaves are smaller and the total area is not increased.

Effect of growth regulators on redistributing dry matter

One of the most important practical effects of growth regulators like CCC or B9 is to redistribute dry matter. These substances shorten stems and divert assimilates to the root system or other underground parts and these grow faster as, for example, in the case of the tuberous root of sweet potato. Some of the consequences of the enlargement of the root system in cereals will be discussed later.

Slowing top growth also causes tubers to develop sooner in Irish potato.[7] The practical advantage of this in a field crop was tested by spraying potato plants with B9, as tubers were forming and again two weeks later. Tubers at first grew more on treated plants but later grew more slowly (Table I). Spraying increased the number of tubers. The final tuber yield was unaffected although the leaf area of treated plants was decreased, confirming that some of the leaf area of field crops is ineffective in producing dry matter.[8] It seems unlikely that B9 would be a suitable growth regulator for potato plants because it increases second-growth of tubers and residues persist in them, but there is evidently further scope for experiments with potatoes; using naturally-occurring inhibitors would be especially interesting.

Table I

Dry weight of tubers sprayed with B9

	Dry wt., g/plant		S.E.
	Untreated	Treated with B9	
30 June	21	24	1·0
25 August	191	204	3·9
4 October	259	257	6·1

Effect of CCC on cereals

At present CCC is the growth retardant most widely used in agriculture. Since it was first described in 1960, there has been widespread interest in its effects and many papers have been published describing its effects on cereals. These have been reviewed by Humphries.[9] As well as shortening wheat stems CCC strengthens them by thickening the lower internodes and so lessens lodging-losses which may be considerable under some conditions. Increase in stem diameter by 11% has been reported, an effect due to thicker cell walls. The shortening of stems depends on the amount of CCC applied, the variety of wheat, the soil moisture and the nitrogen supply; it may be as much as 40%.

At Rothamsted losses from lodging are usually not large and it has not been possible to confirm that CCC prevents them, but many experiments in other parts of England and on the Continent have shown that yields are increased when lodging is prevented by CCC; sometimes more nitrogen can be used without risk of lodging and yields are thereby further increased.

In several experiments at Rothamsted and Woburn,[10,11] however, it has been found that CCC increases yield of crops that did not lodge. Results from these experiments are summarised in Table II.

Grain yield per acre is made up of the following components: number of grains per ear, weight per grain and number of ears per acre. CCC usually increases grains per ear and sometimes ears per acre, but may make the grains smaller, especially of spring

Table II

Examples of increased yields of wheat treated with CCC

Variety	Year	Yields, cwt/acre	
		Untreated*	Treated with CCC
'Phoebus'	1964	39·7	41·9
'Kloka'	1966	37·5	43·5
'Kloka'	1966	31·8	33·8
'Champlein'	1967	54·9	59·0
'Cappelle'	1967	41·5	46·4

* There was no lodging on the control plants.

wheat. Others too have found CCC increases grains per ear. Perhaps CCC slows development slightly so that more grains develop, but this phenomenon needs more detailed investigation.

When CCC increases the number of ears it is because fewer tillers die as the crop grows.[10] Cereal crops always lose many shoots because of competition, and CCC makes the loss smaller (Table III).

Table III

Effect of CCC on shoots of spring wheat

Time	Number of shoots per m^2	
	Untreated	Treated with CCC
14 May	949	934
3 June	753	788
24 June	569	583
24 August	482	550

Small amounts of CCC may occur in the grain of treated crops but with the amounts now recommended the residue is likely to be negligible. However, in experiments where $2\frac{1}{2}$ lb/acre was applied, grain contained enough to be detected by the straight growth test on coleoptiles from grains of treated plants. Dark-grown coleoptiles from grain of treated and untreated plants were floated on water or solutions of IAA (indole-acetic acid) or GA_3 (Gibberellin A_3). Table IVa shows that growth of coleoptiles from grains of CCC-treated plants was inhibited in water, IAA and GA_3; when the two growth hormones were combined the growth rate was equal to that of coleoptiles from untreated plants. Grain contained less CCC when crops were sprayed early than when sprayed late. Table IVb compares the percentage elongation of coleoptiles from grains of plants treated at different times with that of grains from untreated plants.

Not all these experiments show increased yield with CCC, and apparently an increase occurred only when there was a short dry spell at about the time ears emerged. This possibility was tested in an irrigation experiment at Woburn Experimental Station, where 2 in of irrigation water increased grain yield by 10 cwt/acre whereas with CCC without irrigation the increase in yield was only 6 cwt/acre.[11]

Table IV

a. *Growth of 10 mm coleoptile sections from grain of treated wheat*

	Untreated	Treated with CCC
Water	18·6	16·5
GA	21·0	20·0
IAA	23·9	21·9
GA+IAA	23·4	23·3

b. % *elongation of coleoptiles from plants sprayed at different times*

Untreated	100%
5 May	95·5%
27 May	94·2%
6 June	93·8%

It is considered that CCC increases yield because it enlarges the root system, which enables more shoots to survive short periods of drought, especially around the time ears are emerging. Mean weight of roots recovered by pulling is shown in Table V. Plants treated with CCC produce roots that penetrate to deeper layers of soil[12] than do roots of untreated plants. CCC only increased yield when soil moisture was limiting at critical times in the growth of the crop.

Table V

Wheat roots recovered by pulling between May and August (*var. 'Opal', 1965*)

	Mean dry weight, g/m^2
Untreated	54·6
Treated with CCC	65·7

CCC has much less effect on other cereals, i.e. barley, oats and rye, than on wheat. The reason for this is still in doubt; some workers consider it is because CCC does not penetrate readily into the plants or because it is rapidly metabolised and transferred as an inactive quaternary base to the root system. The fact that barley is shortened soon after CCC is applied but later the treated plants grow faster than untreated plants, perhaps supports the idea that it is metabolised. Fig. 1 shows that the lower internodes of barley were shortened by CCC but the upper ones and peduncle were longer than on untreated plants, and the final height was unaffected.

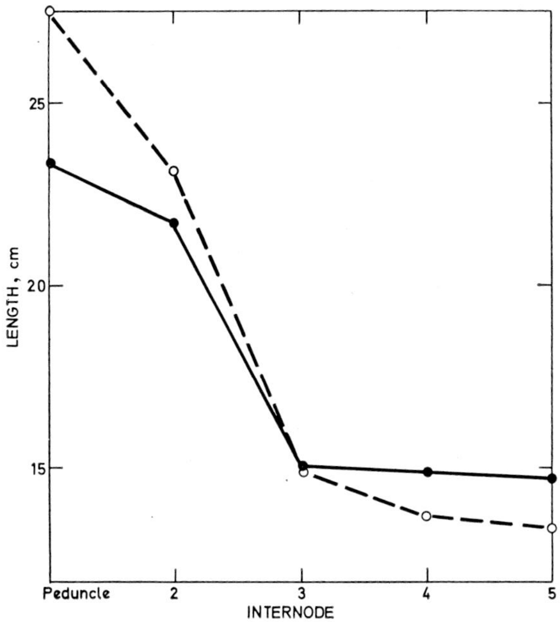

FIG. 1. *Lengths of peduncle and successive internodes of barley at harvest*
● — ● control ○ — — — ○ treated with CCC

Conclusion

This brief account of some experiments with growth regulators applied to field crops shows that yields can be influenced and there is other evidence, not discussed here, that pests and diseases may also be affected by treating plants with growth regulators. Growth regulators can be expected to play an increasing part in agriculture.

References

[1] Tanimoto, T., & Nickell, L. G., Proc. 25th Annual Meeting Hawaiian Sugar Technologists, 1966, p. 184
[2] Humphries, E. C., & French, S. A. W., *Ann. appl. Biol.*, 1960, **48,** 177
[3] Humphries, E. C., & French, S. A. W., *Ann. appl. Biol.*, 1963, **52,** 149
[4] Humphries, E. C., *Ann. Bot.*, 1963, **27,** 517
[5] Humphries, E. C., *Ann. Bot.*, 1968 (in press)
[6] Humphries, E. C., & French, S. A. W., *Ann. appl. Biol.*, 1965, **55,** 159
[7] Dyson, P. W., *J. Sci. Fd Agric.*, 1965, **16,** 542
[8] Humphries, E. C., & Dyson, P. W., *Eur. Potato J.*, 1967, **10,** 116
[9] Humphries, E. C., *Fld. Crop Abstr.*, 1968, **21,** 91
[10] Humphries, E. C., Welbank, P. J., & Witts, K. J., *Ann. appl. Biol.*, 1965, **56,** 351
[11] Humphries, E. C., Welbank, P. J., & Williams, E. D., *Nature, Lond.*, 1967, **215,** 782
[12] *Rep. Rothamsted exp. Stn* 1967, 1968, p. 102

DISCUSSION

Prof. B. B. Stowe.—Was 'Phosphon D' used in any of your experiments, and how did its effects differ?

Dr. Humphries.—'Phosphon D' behaved very similarly to B9 in some pot trials, but because it persists in the soil we did not use it in field experiments.

Dr. V. S. R. Das.—Do fertilisers other than nitrogen have any influence with respect to the action of CCC on crop yields?

Dr. Humphries.—I have not observed any effects, but not many experiments have been carried out on this aspect.